Pocket Chart
Math Activities

Written by
Mary Kurth

Illustrator:
Catherine Yuh

Editor:
Joel Kupperstein

Project Director:
Carolea Williams

D1299612

CTP ©1996, Creative Teaching Press, Inc., Cypress, CA 90630
Reproduction of activities in any manner for use in the classroom and not for commercial sale is permissible.
Reproduction of these materials for an entire school or for a school system is strictly prohibited.

 **able of Contents**

Introduction

Pocket Chart Math Activities is a guide for using a pocket chart to teach basic math concepts. Each chapter reinforces a specific math concept and contains an introductory paragraph, discussion topics, several pocket chart activities, extension activities, a list of related literature, and reproducible pages.

Students learn best when math concepts are introduced using concrete, hands-on manipulatives before they are asked to work with them in the abstract, using paper and pencil. The pocket chart activities in this resource are an ideal pictorial connection between concrete and abstract learning because they incorporate paper versions of common manipulatives and frequently include written follow-up exercises. In addition, nearly all extension activities employ manipulatives to reinforce concepts covered in the chapter.

As a teaching tool, a pocket chart helps introduce, share, and display learning. Its structure and mobility make it convenient for students and teachers alike. Young children enjoy experiencing math with a pocket chart because they can share the learning experience with those around them. Through repeated exposure to pocket chart activities, children develop strong mathematical language skills and improve their ability to count, add, subtract, measure, estimate, graph data, and more.

Although activities can be implemented with large student groups, you may wish to teach them to smaller groups whenever possible. Small group interactions involve each child more often and help you better assess students' levels of understanding.

The amount of time you spend on each math concept depends on students' needs and your curriculum goals. After completing a chapter, keep the pocket chart materials handy so students can practice activities on their own or in small groups. Refer frequently to previous math activities to check for retention of skills and information.

The chapters in this resource may be presented in any order. The book begins with basic math skills such as patterning and counting and ends with more challenging concepts. As you progress through the book, you will find skills often interwoven. For example, as children learn graphing, they are also counting, sorting, classifying, and adding. Introduce or reteach activities from other chapters when appropriate.

Children benefit most by becoming actively involved with math activities presented in a pocket chart. The more children participate, the more success they experience. To involve students even when it is someone else's turn at the pocket chart, have them complete exercises independently using clipboards or small chalkboards. This not only encourages student involvement, but also provides you with a quick, informal assessment method. Ask students to hold up their work so you can see how well they understand the math concept.

The activities that follow are easy to implement, and the required materials inexpensive and readily available. As you incorporate *Pocket Chart Math Activities* into your curriculum, experiment with ideas and activities. Pick and choose activities that work best for you in your learning environment, keeping in mind the specific needs of your students.

Presenting Math Concepts

Place the pocket chart in an area of the classroom where children can gather comfortably on the floor. Be sure the chart is low enough so they can interact with each pocket. Place a table, bookshelf, or basket near the pocket chart for storing word, number, and picture cards; sentence strips; and other materials required for each activity. Keep a pointer, index cards, sentence strips, and a broad-tipped marker on hand for spontaneous math activities.

After presenting each activity, store materials in large resealable plastic bags near the pocket chart. Children enjoy repeating the activities during free time.

Discussion Starters

Introduce each math concept by discussing the questions and topics listed in the *Discussion Starters* section of each chapter. These questions and topics motivate students and encourage them to think about the concept being presented. They prepare students for activities and encourage participation by offering opportunities to relate to or make predictions about the math concept.

Pocket Chart Activities

Each chapter includes detailed instructions for four or more pocket chart activities that focus on the math concept. Many activities call for reproducible picture, number, and word cards. To make these cards more durable, glue them to tagboard or index cards and laminate or cover with clear self-adhesive paper.

Extension Activities

Extension activities offer hands-on, creative-learning, and recording activities that further enhance understanding of math concepts. Some extension activities involve individualized learning while others include the whole group. Many of these activities incorporate the pocket chart. When possible, conduct extension activities near the pocket chart.

Literature Connections

Enrich the study of each math concept with a collection of literature titles. Share these selections with students to expand their understanding of concepts and promote math language development. You may wish to prepare word cards, sentence strips, and picture cards to correspond with a literature selection and use them in the pocket chart to illustrate the connection between math and language arts.

The ability to recognize patterns is a valuable problem-solving tool and a key to mathematical thinking. As students become familiar with patterning, they develop a basis for recognizing sequence and order within our world. The following pocket chart activities offer children a variety of opportunities to recognize, interpret, reproduce, create, and extend patterns. As children participate in these experiences, keep in mind their specific needs and adapt activities accordingly.

PATTERNS

DISCUSSION STARTERS

- What is a pattern?
- Where are patterns in our classroom?
- Would someone like to show and tell about a pattern on their clothing?
- I will say a pattern (such as *red, blue, red, blue . . .*). When you understand the pattern, join in and say it with me.
- I'm going to show a movement pattern (such as *clap, stomp, clap, stomp . . .*). When you understand the pattern, join in by showing the movement pattern.

Pattern Block Copy and Extend

Materials
paper pattern blocks (page 9)
pattern blocks

Directions
1. Reproduce, color, and cut apart several sheets of paper pattern blocks.

2. Model patterns using pattern blocks.

3. Model patterns in the pocket chart using the paper pattern blocks.

4. Begin a pattern in the pocket chart, and invite a student to extend the pattern to the end of the row.

5. Invite pairs of students to repeat steps 2–4. Have one child begin a pattern and the other extend the pattern to the end of the row.

6. When each row holds a pattern, discuss the patterns as a class.

7. Continue the activity until all students have begun or extended a pattern.

Linking Cube Pattern Pass

Materials
6–8 stacks of 10 linking cubes, connected in patterns
bell
3" paper squares (matching cube colors)

Directions
1. Have students sit in a circle. Pass a stack of patterned cubes around the circle. Encourage children to note the pattern as the stack is passed to them.

2. Ring the bell. Have the student holding the stack use paper squares to reproduce the pattern in the pocket chart, then place the stack in the center of the circle.

3. Replay the game, passing a new patterned stack around the circle.

4. When the pocket chart is full, compare patterns and clear each row. Continue the game until all students have had a turn.

Pattern Models

Materials
4" x 6" index cards
marker
paper pattern blocks (page 9)
sentence strips
glue

Directions
1. Print a variety of pattern models on index cards. (For example, print *AB*, *ABC*, *ABB*, or *AABC*.)

2. Prepare pattern strips by gluing colored paper pattern blocks to sentence strips.

3. Place several pattern strips in the chart and model how to interpret each pattern. Then match pattern strips with corresponding pattern models. Provide each child with a pattern model or pattern strip.

4. Invite each child with a pattern model to find a classmate with a matching pattern strip and place both cards in the chart.

5. Play until all pattern models are matched with pattern strips in the chart.

Head, Shoulders, Clap

Materials
5" x 7" index cards
markers
pattern model cards from *Pattern Models* activity

Directions
1. Prepare "action cards" by writing or drawing pictures of the following on index cards: *touch your head, touch your shoulders, touch your knees, stomp your foot, clap*. Make multiple copies of each card.

2. Place pattern model cards in the top row of the pocket chart.

3. Place action cards in the bottom row of the pocket chart.

4. Have a student place a pattern model card in the center of the chart. Have another student place action cards in the corresponding pattern near the model card. Demonstrate the pattern for the class before inviting children to join in.

5. Replace cards and invite students to repeat the activity using other patterns.

Variation
Gather rhythm instruments such as rhythm sticks, bells, triangles, and drums. Draw a picture of each on an index card. Invite groups of children to use pattern model cards and instruments to make musical patterns.

EXTENSION ACTIVITIES

Manipulative Patterns

Provide children with collections of small manipulatives such as buttons, shells, pasta, keys, and bread tabs. Use pattern model cards to create corresponding patterns with manipulatives. Also, invite students to search for patterns on Creative Teaching Press's *Buttons Buttons* floor puzzle (CTP 4204).

Sock Collection

Invite children to bring old or mismatched socks for the classroom sock collection. Store the collection in a basket or box. Have children create sock patterns and record them by drawing pictures on 6" x 18" paper strips. Print corresponding pattern words underneath children's drawings.

Edible Patterns

Provide an assortment of small edibles such as cereal, crackers, raisins, sunflower seeds, and chocolate chips. Divide children into small groups, having one child create a pattern and others copy it. Invite groups to eat their patterns and begin again with a new pattern.

Pattern Headband

Prepare pattern headbands by gluing paper pattern blocks to 2" x 18" tagboard strips. Invite children to work with partners at the pocket chart. Have one child put on a pattern headband and the other reproduce the pattern in the pocket chart. Have children take turns wearing headbands and reproducing patterns.

LITERATURE

A My Name Is Alice by Jane Bayer (Dial)

Buttons Buttons by Rozanne Lanczak Williams (Creative Teaching Press)

Fire Engine Shapes by Bruce McMillan (Lothrop)

I See Patterns by Linda Benton (Creative Teaching Press)

Look Again by Tana Hoban (Macmillan)

Look Around by Leonard Everett Fisher (Viking Press)

Mr. Noisy's Book of Patterns by Rozanne Lanczak Williams (Creative Teaching Press)

Pancakes, Crackers, and Pizza by Marjorie Eberts and Margaret Gesler (Childrens Press)

Paper Pattern Blocks

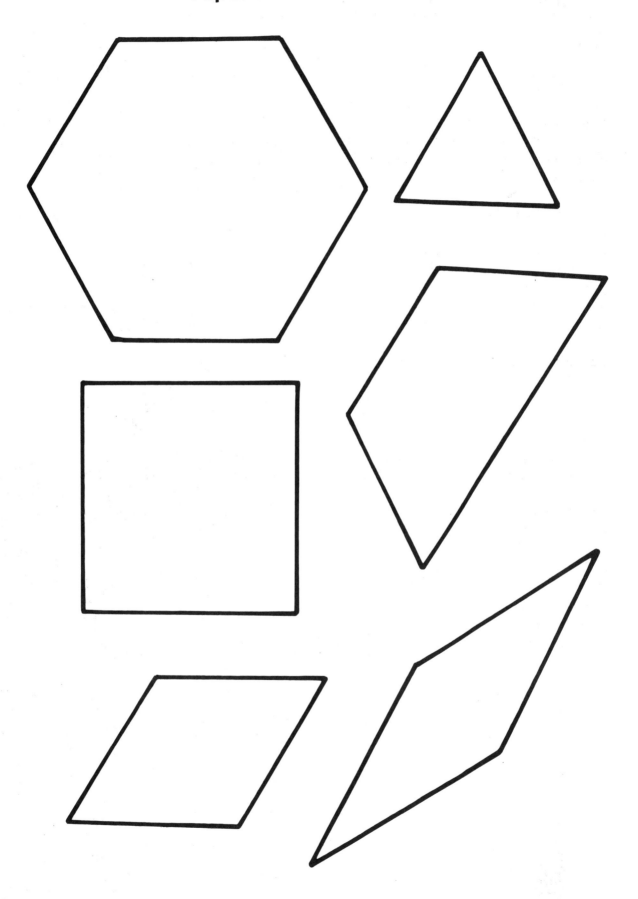

The ability to recognize, label, and form geometric shapes helps children understand the attributes of each shape. The pocket chart activities that follow provide students with many opportunities to learn geometric shapes and terms. Children develop an understanding of shape, size, congruence, and similarity as they become actively involved in these activities.

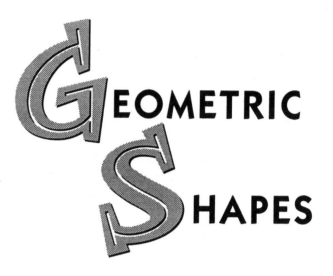

Geometric Shapes

DISCUSSION STARTERS

- Which shapes can you name?
- Where is a circle (square, triangle, rectangle, diamond, oval) in the classroom?
- What shape is our window (door, table, clock)?
- Who would like to draw a square (circle, triangle, rectangle, diamond, oval) on the chalkboard?
- What shapes can we make with our bodies?

Search and Tell

Materials

paper shapes (page 14)

Directions

1. Reproduce and cut apart several sheets of paper shapes. Scatter the shapes around the classroom before children arrive.

2. As children come into the room, invite them to search for shapes and bring one each to the pocket chart area.

3. When all shapes are found, invite children to place them in the pocket chart, one by one. Have each child name his or her shape and tell three facts about it. (For example, *This rectangle has four sides. Two sides are long. It looks like a door.*)

4. After all shapes are in the pocket chart, remove them, one by one, as children say each shape's name.

5. Repeat the activity, encouraging children to find new shapes.

Shapes Walk

Materials

crayons or markers
4" x 6" index cards

Directions

1. Take students on a "shapes walk." Have children identify shapes in and around the school grounds.

2. Back in the classroom, discuss shapes students observed. Distribute crayons and index cards, and have each child draw a picture of a shape observed during the walk. (If possible, have each child draw a different picture.)

3. Print words on index cards to match students' pictures.

4. Invite each child to place his or her picture and word card in the pocket chart and share aloud.

5. After all children have placed their cards in the chart, remove only the word cards and play a group matching game. Place one word card in the bottom of the chart. Ask the child who drew the matching picture to place the word card near its corresponding picture card. Continue until all cards are matched.

Variation

Print sentences on sentence strips to match each picture card. (For example, *Madeline saw a clock that was a circle. Joseph saw a wooden door that was a rectangle.*) Place a picture card and sentence strip in each row of the pocket chart. Read the chart aloud, pointing to the words as you read. Invite children to read aloud from the chart. Clear the chart and repeat the activity with the rest of the picture cards and sentence strips.

Stand Up

Materials

scissors
construction paper
paper shapes (page 14)

Directions

1. Cut six large construction-paper shapes to match those on page 14.

2. Have children gather on the floor near the pocket chart.

3. Provide each child with a reproducible paper shape. Have children with like shapes sit together.

4. Place the construction-paper square in the pocket chart and call out, *Squares, stand up!*

5. Remove the square and call out new directions to other groups. For example, call, *Circles, sit down. Rectangles, kneel. Triangles, put your hands on your heads.*

Variation

Incorporate this activity into a unit on body parts. Give commands such as *Squares, put your shapes on your thighs. Triangles, put your shapes on your wrists. Ovals, put your shapes on your backs.* Each time a group is given a direction, place the shape in the pocket chart.

Shapes Sort

Materials

scissors
shape pictures (page 15)
construction paper
crayons or markers

Directions

1. In advance, cut apart three different shape pictures for each student.

2. Cut six large construction-paper shapes (matching those on page 14) and place them in the top row of the pocket chart.

3. Have children color their shape pictures.

4. Invite children to place their pictures in the column beneath the corresponding shape. Encourage them to name each shape and identify each picture.

5. After all shape pictures have been sorted, invite children to choose three new pictures from the chart and repeat the sorting activity.

EXTENSION ACTIVITIES

What Am I?

Glue each shape (page 14) to a 5" x 7" index card. On the back of each card, print sentences about the shape. (For example, *I am round. I am the shape of the sun. I have no corners. What am I?*) Place all cards shape-side-down in the pocket chart. Read one card aloud, and choose a child to name the shape then check the answer by turning over the card. Play until all shapes have been guessed.

Shape Sorting

Cut pictures of shapes from magazines. Glue each to a 3" x 5" index card and laminate for durability. Staple six envelopes along the bottom of the pocket chart. Draw one shape on each envelope. While working in pairs, have one child place pictures in the chart in random order and the other remove each card from the chart, name the shapes, and place them in corresponding envelopes.

Copy-Cat Yarn Shapes

Place six large construction-paper shapes in the pocket chart. Give each child in a small group an 18" loop of yarn. Have one student place a shape in the pocket chart, name it, and form it with yarn. Have the rest of the group form and name the shape as well. Have new leaders form shapes until all six have been formed.

Shape Word Match

Print shape words (*triangle, circle, square, rectangle, diamond, oval*) on 5" x 7" index cards to correspond with each shape on page 14. Color, cut out, and glue each shape to a 5" x 7" index card. Place the shapes in the pocket chart. Hand out the word cards. Invite children to match word cards and shapes.

Shape Sorting Tray

Gather small boxes and an assortment of small objects in six shapes. Draw one of the six shapes on the side of each box. Place the collection at a table and invite students to sort items into boxes, naming shapes as they work. Have them tell a friend about the shapes they sorted.

LITERATURE

Animal Shapes by Brian Wildsmith (Oxford)

Color Zoo by Lois Ehlert (HarperCollins)

Grandfather Tang's Story by Ann Tompert (Crown)

I See Shapes by Marcia Fries (Creative Teaching Press)

Shapes by John Reiss (Macmillan)

Shapes Game by Stan Tucker (Holt)

Shapes in Nature by Judy Feldman (Childrens Press)

Shapes, Shapes, Shapes by Tana Hoban (Greenwillow)

Paper Shapes

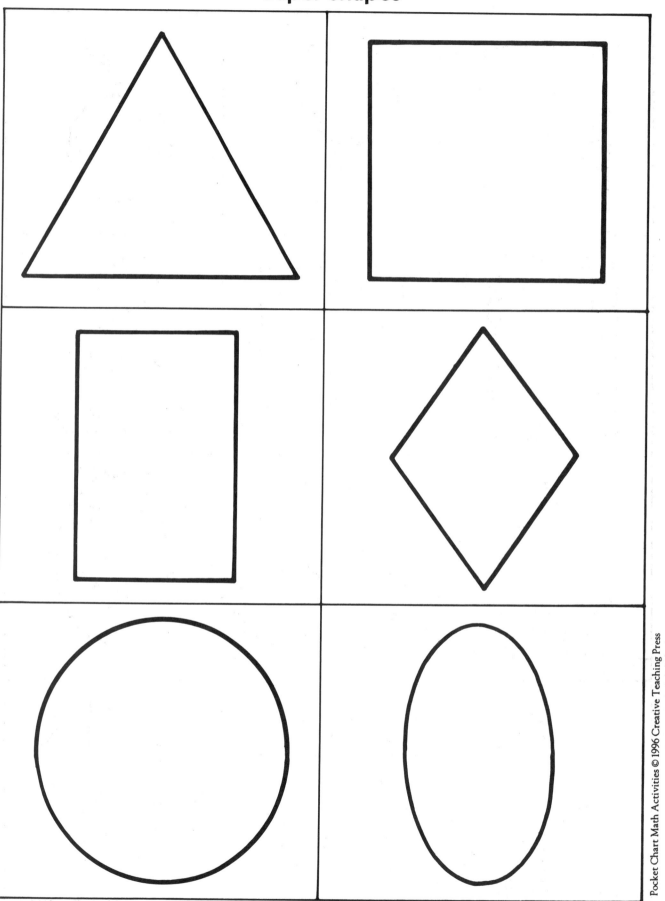

Pocket Chart Math Activities © 1996 Creative Teaching Press

Shape Pictures

Pocket Chart Math Activities ©1996 Creative Teaching Press

Young children need many opportunities to count in order to develop number sense and an understanding of numerical order. In this chapter you will find a variety of activities that help students learn skills such as rote counting, one-to-one correspondence, conservation of number (the fact that size and arrangement of objects does not affect quantity), counting on, and counting backward. With repeated opportunities to count in a variety of ways, students establish the link between rote counting and other number concepts.

Counting

DISCUSSION STARTERS

- Turn to a partner and count to 20.
- Who would like to come to the pocket chart and count these shapes for me?
- Count backward from ten.
- I will count aloud, then stop. Continue counting from where I leave off.

Point and Count

Materials

crayons or markers
scissors
paper bears (page 20)

Directions

1. Color and cut out paper bears.

2. Place a different number of bears in each row of the pocket chart.

3. Designate each row in the pocket chart as *row one*, *row two*, and so on.

4. Invite a child or group of children to point to and count the bears in the top row.

5. Have different children count each row.

6. Clear the chart and place new numbers of bears in the rows. Play until all children have had a chance to point and count.

Variation

Give each student a paper bear. Have students sit in a circle and count off around the circle. Invite each child to place his or her bear in the chart and count on from the person before. (The first child says, *one*, the second says, *two*, the third, *three*, and so on, until all bears are counted.)

Caterpillar

Materials

2 shoe boxes
2" paper circles
2 dice
marker

Directions

1. Divide students into two teams. Designate one team the "top row team" and the other the "bottom row team." Have each team sit in a line facing the pocket chart. Place a shoe box containing 25 paper circles and a die at the front of each line. Draw caterpillar faces on two circles, and place one in the top row and one in the bottom row of the chart.

2. Instruct the first person in each line to roll a die, count out the corresponding number of circles, and place them in the chart.

3. Have children continue taking turns rolling dice, counting, and adding circles to their team's caterpillar.

4. After all students have taken turns, invite a member of each team to lead the team in counting the circles in their caterpillar.

Note: If caterpillars grow too big for their rows, curl their "tails" toward the opposite end of the pocket chart.

Pass and Count

Materials

8–10 sets of counters (different number in each set)
plastic bags
bell
3" x 5" index cards
marker
pushpins

Directions

1. Place each set of counters in a plastic bag.

2. Have children sit in a circle. Give a bag of counters to a student and have him or her pass it around the circle.

3. After a short while, ring the bell. Ask the child holding the bag to spill its contents onto the floor and count the objects aloud.

4. Print the number of objects on an index card. Have the child refill the bag, place the card in the top row of the chart, and pin the bag next to the card.

5. Repeat steps 2–4 with other bags. Play until all bags have been counted and placed in the chart with number cards.

6. Place cards and bags in a shoe box. Make them available for children to count and match during free time.

Counting On

Materials

number cards (numbers *0–9* on index cards)
bell
3" x 5" index cards
marker

Directions

1. Give each child a number card.

2. Invite children, one by one, to place cards in the pocket chart and name the numbers.

3. Each time a child places a number in the chart, have the class count on from that number. Ring the bell to end counting. For example, if a student puts a seven in the chart, the class begins with eight and counts until the bell rings. Print the highest number counted on an index card and place it in the same row as the first to show how high the class counted.

4. Repeat the activity until all children have placed cards in the chart.

Counting Down

Materials

paper bears (page 20)

Directions

1. Place several bears in the pocket chart.
2. Point to each bear as students count aloud.
3. Count down as you remove each bear from the chart.
4. Repeat the activity with different numbers of bears for children to count.

EXTENSION ACTIVITIES

Counting Bears
Place paper bears (page 20) in a shoe box with several number cards. Instruct children to count bears to match each card. Encourage children to have friends check their work. For more counting practice, invite students to work with the *Spiders, Spiders Everywhere!* floor puzzle published by Creative Teaching Press (CTP 4212).

Counting Objects from Home
Send a backpack or book bag home with a different child each day. Invite students to bring it back filled with objects for the class to count. The next day, ask the designated child to remove the objects and count them aloud for the class. Print sentence strips to correspond with the number of objects the child brings. For example, *Sophie brought six small dolls. Sophie brought fifteen barrettes. Sophie brought ten plastic animals.* Leave objects and sentence strips near the pocket chart for children to count and read during free time.

Counting Jars
Place three jars of manipulatives at a table (a different manipulative in each jar). Invite children to count the objects in each jar. Have children record their counting by drawing pictures of the jars and manipulatives and writing each number.

LITERATURE

A-Counting We Will Go by Rozanne Lanczak Williams (Creative Teaching Press)

Counting Wildflowers by Bruce McMillan (Lothrop)

The Crayola® Counting Book by Rozanne Lanczak Williams (Creative Teaching Press)

Each Orange Had Eight Slices by Paul Giganti (Greenwillow)

Five Little Monsters by Rozanne Lanczak Williams (Creative Teaching Press)

How Many? by Rozanne Lanczak Williams (Creative Teaching Press)

Mouse Count by Ellen Stoll Walsh (Harcourt Brace Jovanovich)

One Hunter by Pat Hutchins (Mulberry)

Rooster's Off to See the World by Eric Carle (Picture Book Studio)

What Do You See? by Rozanne Lanczak Williams (Creative Teaching Press)

Paper Bears

Pocket Chart Math Activities © 1996 Creative Teaching Press

When children are comfortable with basic counting skills such as rote counting, one-to-one correspondence, and conservation of number, they are ready to move on to the more challenging concept of skip counting. Skip counting helps students estimate, organize information, and improve their speed of computation. Students also gain confidence in speaking and counting in front of a group and in their ability to manipulate and count math materials. Working with the following activities helps students learn to skip count by twos, fives, and tens.

DISCUSSION STARTERS

- How many ways can you count to 20? to 100?
- What comes in groups of two? five? ten?
- Who can count these blocks by twos, fives, and tens?
- How is skip counting helpful?

Feet, Hands, Fingers

Materials

handprints (page 25)

Directions

1. Reproduce, color, and cut apart several hand-prints.

2. Invite children to sit in front of the pocket chart, extending their legs forward. Go around the group and count feet by twos.

3. Repeat step 2, counting hands instead of feet.

4. Place several pairs of paper handprints in the pocket chart.

5. Invite students to count by twos as you point to pairs of handprints.

6. Add more handprints and count again. Continue adding handprints as the class counts by twos.

Variation

Explain that there are five fingers on each hand and ten fingers on each pair of hands. Repeat the activity counting fingers by fives and tens.

Counting by Fives

Materials

linking cubes
paper linking cubes (page 26)
crayons or markers

Directions

1. Have children sit in a circle. Place a large pile of linking cubes in the center.

2. Invite children to connect stacks of five linking cubes in front of them. Walk around the circle and, as a group, count the stacks by fives.

3. Invite children to color and arrange paper linking cubes to match their stacks.

4. Place several paper "stacks" in the pocket chart. Have the class count by fives as you point to the stacks.

Variation

Have children draw sets of five tally marks on index cards and place cards in the pocket chart. Invite children to skip count the tally marks as you point to each set.

Skip Count Passing Game

Materials
linking cubes
handprints (page 25)

Directions

1. Connect a stack of two, a stack of five, and a stack of ten linking cubes.

2. Have children sit in a circle.

3. Place 20 handprints in the pocket chart.

4. Sing "The Skip Count Song" for twos, and have children pass the stack of two cubes around the circle. When the song ends, have the student holding the stack count the hands in the pocket chart by twos.

5. Repeat the activity, singing "The Skip Count Song" for fives and tens. Have children left holding the five and ten stacks count fingers by fives or tens.

Skip count, skip count, Count by twos.

Skip count, skip count, Count by twos.

Skip count, skip count, Count by twos.

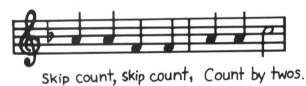

We can count to twen – ty!

Skip Counting Cards

Materials
5" x 7" index cards
glue
manipulatives (buttons, toothpicks, blocks)
crayons or markers

Directions

1. Give each student three index cards, glue, and a collection of manipulatives.

2. Have students glue two objects to one card, five to another, and ten to the third. Instruct them to print the number of objects on the back of each card.

3. Invite students to place their "two" cards in the pocket chart, and count the total number of cards. Record the results.

4. Skip count the objects, card by card, as a class.

5. Repeat for "five" and "ten" cards.

EXTENSION ACTIVITIES

Counting Partners

Divide the class into pairs. Invite one partner to fill the pocket chart with a variety of felt shapes while the other counts a category of shapes (red circles, blue shapes, squares). Have children clear the chart, trade places, and repeat the activity. Incorporate skip counting as children place shapes in the chart in groups of two, five, or ten.

Handfuls

Have children sit in pairs. Provide each pair with a set of manipulatives (blocks, linking cubes, buttons, paper clips). Instruct one partner to grab a handful of objects and separate them into groups of two, five, or ten, and the other to skip count the objects. Have partners switch roles and repeat the activity.

Chalk Drawings

Take students to a sidewalk or blacktop area. Using colored chalk, have students draw a set of pictures that can be counted by twos, fives, or tens. When students finish, invite the class to walk from set to set, having the student who drew the pictures lead the class in skip counting the set.

LITERATURE

Count and See by Tana Hoban (Macmillan)

Just Cats: Learning Groups by John Burningham (Viking)

Numbers by Sara Lee Anderson (Dutton)

The Skip Count Song by Rozanne Lanczak Williams (Creative Teaching Press)

Two Ways to Count to Ten by Ruby Dee (Holt)

What Comes in Threes? by Marlene Beierle (Creative Teaching Press)

What Comes in 2s, 3s, and 4s by Suzanne Aker (Simon & Schuster)

Handprints

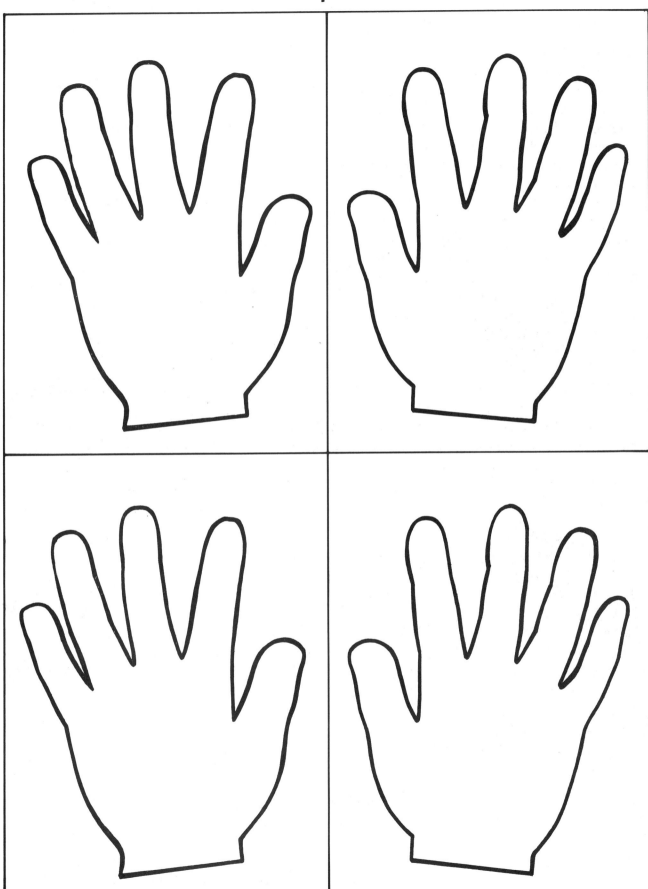

Pocket Chart Math Activities © 1996 Creative Teaching Press

Paper Linking Cubes

Pocket Chart Math Activities © 1996 Creative Teaching Press

Sorting and classifying groups of objects are math skills that help young children organize information and see relationships. As students participate in the following pocket chart activities, they improve their ability to think critically, make predictions and comparisons, draw conclusions, and express ideas. Students learn to recognize similarities and differences when given opportunities to sort, classify, and solve problems.

SORTING & CLASSIFYING

DISCUSSION STARTERS

- How are children in our class alike?
- How are children in our class different?
- Could we form groups of children in our class that are alike in some way?
- (Show attribute blocks.) How are these blocks alike?
- (Organize blocks by size, shape, or color.) How have I grouped these blocks?

Attribute Block Sort

Materials

attribute blocks
paper attribute blocks (page 31)

Directions

1. Sort attribute blocks by size, shape, thickness, or color. Invite children to guess how you are sorting.

2. Sort paper blocks in the pocket chart. Tell the class how you are sorting. Hand a paper block to each child.

3. Invite children to place their blocks in the chart next to those with similar attributes.

4. Repeat the activity, sorting by a different attribute. This time, do not tell children how you are sorting. Encourage them to determine how you are sorting before they place shapes in the chart.

5. Leave materials near the pocket chart for children to explore during free time.

Variation

Print word cards for each attribute. Place these in the chart near the corresponding groups after they have been sorted.

Playing Card Sort

Materials

playing cards

Directions

1. Gather children into groups of four or five. Model ways to sort playing cards in the pocket chart (by color, number, or suit).

2. Give each group 15–20 playing cards, and have children decide how to sort their cards.

3. When sorting is complete, invite groups to come to the pocket chart to show and tell how they sorted their cards.

4. Have groups select different attributes and sort cards again.

Variation

As each group finishes sorting, invite the class to determine how cards have been sorted.

Letter/Sound Sort

Materials
3" x 5" index cards
marker
pictures (from magazines or photographs)

Directions

1. Print two letters on index cards (one per card).

2. Gather an assortment of pictures that begin with those letters.

3. Place the letter cards in the top row of the pocket chart. Provide each child with a picture.

4. Say to children, *If you have a picture that begins with _____, please put it in the pocket chart.* Supervise children as they place pictures beneath corresponding letters.

5. Repeat until all pictures are sorted.

Variation

Print simple word cards for children to read and sort. Have them sort words by beginning letters as they did the picture cards.

Button Sort

Materials
buttons
sentence strips
marker

Directions

1. Have each group sort a collection of buttons and tell about their categories.

2. Print a sentence strip explaining how each group sorted. (For example, *One group sorted buttons by size. Another group sorted buttons by color. One group sorted buttons by how many holes they had.*)

3. Place these in the pocket chart and invite children to read the strips aloud.

4. Invite groups to sort buttons in new ways.

Pile Up

Place piles of buttons, blocks, and cotton balls on a table (one pile of each). Invite children to estimate how many are in each pile and form new piles containing the same amounts. After children form three piles, have them count and compare. Repeat this activity, incorporating different sizes and amounts of manipulatives in the piles. For more practice, invite students to work with the *I See Colors* floor puzzle by Creative Teaching Press (CTP 4207). Have them estimate and count the number of colored balloons.

Sink or Float?

Gather common classroom objects, a bowl of water, and a towel. Draw each object on a 3" x 5" index card. Place cards reading *sink* and *float* in the top pocket of the chart. Invite children to predict which items sink and which float and place picture cards under appropriate headings. After children experiment with the objects, have them check to see which guesses were correct.

Class Sorting Book

Provide small groups of children with manipulatives to sort (buttons, nuts and bolts, shells, multicolored pasta). Have groups practice sorting in a variety of ways. Then have them record one way they sorted by drawing groups of objects on a sheet of paper. Combine pages into a class book.

Sorting People

As a class, sort students into groups by common attributes such as hair color, hair length, gender, eye color, or clothing color. After selecting an attribute, have students find people who share that attribute and form groups with them. On the chalkboard, write descriptive sentences about the groups, such as *Eleven students are wearing pants. More girls have dark hair than boys.* Repeat the activity several times, each time for a new attribute.

The Button Box by Margaret S. Reid (Dutton)

Buttons Buttons by Rozanne Lanczak Williams (Creative Teaching Press)

Goldilocks and the Three Bears by Jan Brett (Putnam)

Hats, Hats, Hats by Ann Morris (Lothrop)

I See Colors by Rozanne Lanczak Williams (Creative Teaching Press)

Is It Red? Is It Yellow? Is It Blue? by Tana Hoban (Greenwillow)

People by Peter Spier (Doubleday)

Sizes by Jan Pienkowski (Simon & Schuster)

Paper Attribute Blocks

Pocket Chart Math Activities © 1996 Creative Teaching Press

This chapter introduces students to calendar concepts. As young children study the days of the week, months of the year, seasons, and dates, they gain a better understanding of time and duration, number patterns, graphing, and sequencing. Repeated experiences with the calendar reinforce and extend these skills. As students participate in the following activities, encourage them to verbalize often and ask questions. They soon develop calendar vocabulary and a variety of basic math concepts to apply to their daily lives at school and at home.

CALENDAR

DISCUSSION STARTERS

- What do you know about calendars?
- Why is the calendar important?
- What are the days of the week? months of the year? seasons?
- What is today's date?
- What are some things that occur in the month of December (or other months)?

POCKET CHART ACTIVITIES

Days of the Week

Materials

sentence strips
marker
magic wand (pointer)
5" construction-paper star

Directions

1. Prepare sentence strips (one per day) describing events that happen on the days of the week. For example, *On Monday, we go to gym class. On Tuesday, we have chocolate milk at snack time. On Wednesday, we have art class.*

2. Place the strips in the pocket chart in chronological order.

3. Reread the sentence strips aloud with the class, pointing to words as you read.

4. Have a student place the star in the row that corresponds to the current day of the week. Invite him or her to read that sentence aloud, pointing to the words with the magic wand.

Note: Make this part of daily calendar activities. Read the entire chart each day and invite various students to read daily sentence strips. Change the sentences from time to time. If a birthday, special event, or holiday is coming up, include it in the chart.

Missing Letter

Materials

3" x 5" index cards
marker
paper clips
days of the week chart
months of the year chart

Directions

1. Write days of the week and months of the year on index cards, one letter per card. Clip the letters of each word together and write the word in the corner of one card.

2. Using days and months charts, review the days of the week and months of the year.

3. Place letter cards for *Monday* in the pocket chart face down so they will be in order when turned over.

4. Invite children to name letters. As a letter is mentioned, turn over that card. Print each letter on the chalkboard as it is named.

5. Have children raise their hands when they think they know the word. If they guess correctly, turn over all the letters in the chart.

Variation

Use this activity when introducing a specific holiday or season. Children will enjoy calling out letters and guessing seasonal or holiday words.

Season Sort

Materials
5" x 7" index cards
crayons or markers
seasonal activity pictures (page 36)
scissors
glue

Directions
1. Print each season's name on an index card.

2. Color and cut out seasonal activity pictures and glue them to index cards.

3. Place each season name card in a corner of the chart. Read the cards aloud as a class.

4. Hold up one picture card and invite a child to tell about the picture and the season with which it corresponds. Have the child place it in the chart near the matching season name card.

5. Play until all cards are sorted.

6. Have students suggest other activities or events common to each season.

Missing Months

Materials
3" x 5" index cards
marker

Directions
1. Print names of the months on index cards and place them in the pocket chart in chronological order.

2. Invite the class to read each month's name as you point to the cards.

3. Remove one card from the sequence.

4. Go through the months again, having the class clap as you point to each card and say the name of the missing month. Remind children to think of months' names as they clap.

5. Continue removing cards from the sequence as children clap and name missing months.

Variation
Remove two or more cards at a time. Divide the class into groups according to hair color, gender, age, or other categories, and invite one group at a time to read and clap the sequence.

EXTENSION ACTIVITIES

Today Is Monday

Read *Today Is Monday* by Eric Carle. Print word cards to match foods and days of the week in the story. Distribute word cards to 14 children. Reread the story, having children place cards in the chart as they are mentioned. Leave the book and cards near the pocket chart so children can match them during free time.

Pocket Chart Calendar

Use the pocket chart as a calendar. Cut 31 three-inch paper squares. Print numbers *1–31* on the squares in alternating colors. Place word cards for months and days of the week across the top of the pocket chart, and place number cards face down under the days of the week. Each day, turn over one number card, noting the color pattern that appears as days go by. To highlight special days or events, place a 4" x 6" index card behind the given date and print the event's name near the top of the card. Children will enjoy counting down to special events using the pocket chart calendar.

Individual Calendar Grid

Prepare a blank calendar grid to use throughout the year. Each month, duplicate one per child and invite children to fill in the month, days of the week, and dates. Encourage children to draw seasonal decorations at the top of each page and label special events and days. Have students use their calendars to answer questions such as *On what days of the week does the month begin and end? How many Fridays are in this month? How many dates have a six in the ones place?*

Days of the Week Diaries

Provide each child with a seven-page blank book, and have children write a day of the week at the top of each page. Each day, invite children to draw or write about one thing they did on that day. Ask them to read their diaries to a friend at the end of the week. For more practice, have students work with the *All Through the Week with Cat and Dog* floor puzzle by Creative Teaching Press (CTP 4209).

LITERATURE

All Through the Week with Cat and Dog by Rozanne Lanczak Williams (Creative Teaching Press)

Cookie's Week by Cindy Ward (Putnam)

The Four Seasons by Rozanne Lanczak Williams (Creative Teaching Press)

Four Stories for Four Seasons by Tomie DePaola (Prentice-Hall)

The Month Brothers by Samuel Marshak (Morrow)

Through the Year with Harriet by Betsy Maestro (Crown)

Today Is Monday by Eric Carle (Philomel)

The Very Hungry Caterpillar by Eric Carle (Putnam)

Seasonal Activity Pictures

Pocket Chart Math Activities © 1996 Creative Teaching Press

Learning to identify numbers and number words is the foundation on which the understanding of other math concepts is built. Number recognition establishes a link between counting and other math concepts taught in kindergarten and first grade. Although many skills in identifying numbers and number words involve rote learning, the following activities include fun ways to teach these skills using the pocket chart.

Number & Number Word Recognition

DISCUSSION STARTERS

- Where do we have numbers in our classroom?
- (Write a number on the chalkboard.) What number is this?
- (Write number words on the chalkboard.) What do these words say?
- Who can write the number *23* (or other numbers)?
- Why is it important that we learn to recognize numbers?
- Where do you see numbers when you're not in school?

Rhythm Sticks

Materials

rhythm sticks
number cards (numbers *0–9* on index cards)

Directions

1. Have children sit in a circle holding rhythm sticks.

2. Give each child a number card.

3. Have children listen as you chant the following, keeping the beat with rhythm sticks: *If you have a number and its name is* three, *put it in the pocket chart just for me.* Have children holding threes place their cards in the chart and return to the circle.

4. Invite children to play to the beat as you continue the chant asking for new numbers.

5. Have a volunteer arrange cards in numerical order.

Quarters in a dollar

Two plus two

Number of legs on a table

Comes before five

Two-Team Guessing Game

Materials

number cards (numbers *0–9* on index cards)

Directions

1. Divide the class into two groups. Have them sit in two lines facing each other.

2. Place number cards face up in a row between the two teams.

3. Have the first person in each line stand up.

4. Have one person describe a number without saying its name. For example, *It has straight lines, comes before five, and a table usually has this many legs.* The person from the other team has to find the number and place it in the pocket chart. Both children then go to the ends of their lines and the next two stand up.

5. Continue until all children have had a chance to describe and find numbers.

Three of a Kind

Materials

number cards (numbers *1–12* on index cards)
number word cards (page 42)
set cards (page 43)
dice

Directions

1. Provide each student with one or more number, number word, or set cards.

2. Roll a die and name the number of dots showing.

3. Invite children holding those number, number word, or set cards to place their cards together in the pocket chart.

4. After cards one through six are in the chart, add the second die. Continue the game until all cards are in the chart. If a number is rolled a second time, roll again until the dice show a new number.

Variation

Leave the sets of cards near the chart. During free time, invite children to work in pairs. Have one student mix up the cards in the chart and the other sort them back into their groups.

Number Sequence Relay

Materials

number cards (numbers *1–10* on index cards)
yarn

Directions

1. Prepare two sets of number cards. Place a set in random order in the bottom row of each half of the pocket chart.

2. Have children form two lines five feet from the pocket chart.

3. Hang yarn down the center of the chart to divide it in half.

4. On the command *Ready, set, go,* have the first team member go to the chart, find the 1 (from the bottom row), place it in the top left side of his or her half of the pocket chart, and go to the back of the line. Have the next team member place the 2 to the right of the 1.

5. The game continues as teams send children, one by one, to place numbers in proper sequence in the chart. The first team to place all ten numbers in order is the winner.

Variation

Incorporate number sequencing from 11–20 when children are ready. Make number cards to be used in place of, or in addition to, numbers *1–10*.

It's in the Bag

<u>Materials</u>
small objects (beans, marbles, cotton balls)
large clear bag
marker
3" x 5" index cards
pushpins
2" tagboard squares

<u>Directions</u>
1. Place twelve small objects in a bag.

2. Print *12* on an index card. On two other cards, print one higher number and one lower number (e.g., *4* and *24*).

3. Place the cards in the top of the pocket chart.

4. Pass one object from the bag around the circle and have children study its size and shape. Pin the bag next to the pocket chart.

5. Ask, *How many objects do you think are in the bag—4, 12, or 24?*

6. Invite each child to place a tagboard square under the number of objects he or she thinks are in the bag to create a graph of predictions. Display the bag for awhile so children can think about and compare their estimations.

7. Near the end of the day, invite one or two children to count the objects. Discuss the accuracy of students' predictions.

<u>Variation</u>
Use different objects in the bag each time you estimate. Invite children to count objects on their own without giving answers away. As students' estimating skills improve, increase the number of objects in the bag.

Sequencing Strips

<u>Materials</u>
sentence strips
marker

<u>Directions</u>
1. Write five-part number sequences on sentence strips, omitting one number from each strip. For example, *1, 2, 3, __, 5* or *7, 8, __, 10, 11.*

2. Place strips in the chart.

3. As a class, chant the number sequence on each strip, clapping and naming the missing number.

4. Continue until all cards have been chanted.

<u>Variation</u>
Invite children to make their own sequence strips. Have the class chant and clap the new sequences.

EXTENSION ACTIVITIES

Sidewalk Number Groups
Provide each child with chalk and a designated space on the sidewalk or blacktop. Assign each child a number. Invite him or her to draw the number and a corresponding set of objects, and write the number word. When all children are finished, take a walk through the colorfully numbered sidewalk spaces.

Individual Sequencing Game
Place a set of number cards in a lunch bag or envelope. Have children spill out the numbers and put them in order as quickly as possible. (If the pocket chart is available, invite children to arrange numbers in the chart.)

Number Match
Place number and word cards face down in the pocket chart. Invite children to turn over two cards, hoping to find a match. If the cards match, have students keep them. If not, have students turn them back over.

How Many of Us?
Take photographs of groups of children or classroom objects. Glue each photo to tagboard and cover with self-adhesive paper. Print a number word card to match each photo, and invite children to match number words and photographs.

LITERATURE

The Bugs Go Marching by Rozanne Lanczak Williams (Creative Teaching Press)

Crictor by Tomi Ungerer (Harper & Row)

How Many? by Rozanne Lanczak Williams (Creative Teaching Press)

I Know More about Numbers by Dick Bruna (Methuen)

The Right Number of Elephants by Jeff Sheppard (HarperCollins)

Sea Squares by Joy Hulme (Hyperion)

Ten, Nine, Eight by Molly Bang (Greenwillow)

Who Wants One? by Mary Serfozo (Macmillan)

Number Word Cards

one	two	three
four	five	six
seven	eight	nine
ten	eleven	twelve

Pocket Chart Math Activities ©1996 Creative Teaching Press

Set Cards

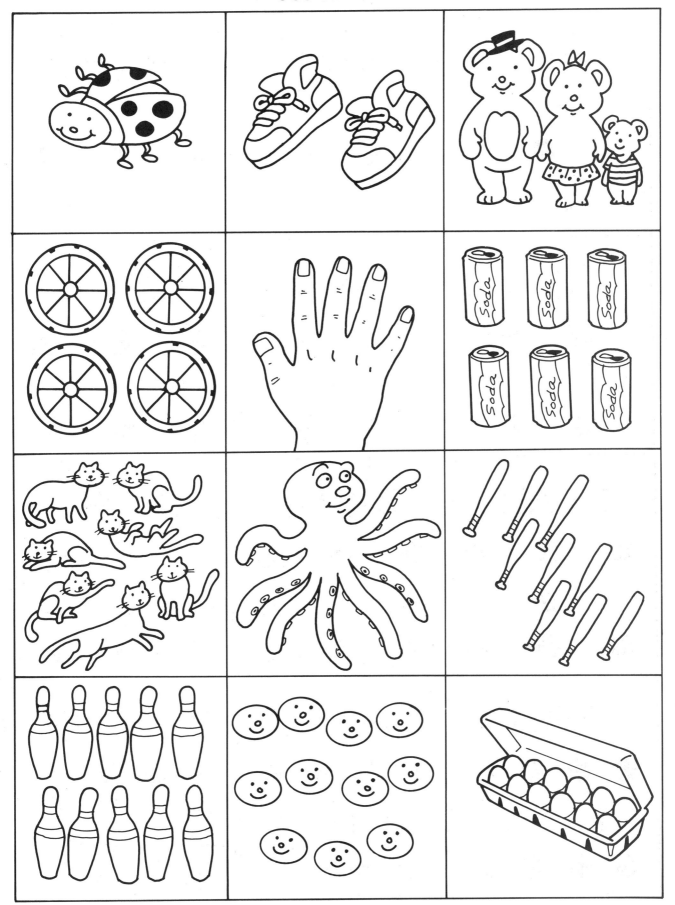

Understanding the concept of place value improves counting skills and prepares students for more advanced arithmetic skills such as addition and subtraction with regrouping. The pocket chart provides a simple, visible way to set up place value columns and manipulatives for students to work with. It is also an excellent focal point for small- or large-group lessons. In the following activities, children count large groups of objects, improve their understanding of the base-ten number system, and observe patterns in numbers larger than ten. Encourage children to use appropriate vocabulary (*digit, ten's place, one's place*) and manipulate objects as often as possible when learning place value.

PLACE VALUE

DISCUSSION STARTERS

- (Write a one-digit number and a two-digit number.) How do these numbers look different?
- What is a digit? What digits are in the number *15*?
- In the number *15*, what does the 1 mean? the 5?
- How many tens in 15? How many ones?
- What is the smallest number with two digits? What is the largest?

POCKET CHART ACTIVITIES

Counting Stars

<u>Materials</u>

stars (page 48)
colored paper
scissors
two 5" x 7" index cards (labeled *tens* and *ones*)
10 envelopes

<u>Directions</u>

1. Copy several pages of stars on colored paper and cut them apart.

2. Place a set of 17 stars in the right side of the pocket chart.

3. Place the *tens* and *ones* cards on the left side of the chart.

4. Count the stars aloud as a class.

5. Count a second time. When you have counted ten stars, place them in an envelope.

6. Place the envelope beneath the *tens* card and the remaining stars beneath the *ones* card.

7. Count the tens and ones. Be sure this number matches the total from step 4.

8. Clear the chart and repeat for a new number.

<u>Variation</u>

Have students record counting individually using star stickers and construction paper.

Linking Cubes

<u>Materials</u>

paper linking cubes (page 26)
scissors
number cards (numbers *0–9* on index cards)
linking cubes

<u>Directions</u>

1. In advance, reproduce and cut apart several pages of paper linking cubes.

2. Invite a child to place two number cards in the pocket chart and read the two-digit number aloud.

3. Have a second student build the number using paper linking cubes. Be sure the student places stacks of cubes (tens) on the left side of the chart and single cubes (ones) on the right. Have other students use linking cubes to build the number individually.

4. Ask a third child to write the number on the chalkboard.

5. Count the paper cubes aloud as a class.

6. Clear the chart and repeat the activity with new students and a new number.

Place Value Relay

Materials

number cards (numbers *0–9* on index cards)

Directions

1. Print three or four sets of number cards.

2. Divide the class into large teams. Give each team a set of number cards. Be sure each team member has at least one card.

3. Call out a two-digit number. Have team members holding the two digits form the number in the pocket chart as quickly as possible. Give three points to the first team finished, two to the second, and one to the third.

4. Have students remove cards from the chart and return to their seats. Repeat steps 2 and 3 until one team reaches 20 points.

Variation

Instead of giving cards to team members, place them in the bottom of the chart. Have teams line up and send students in order to form numbers in the pocket chart.

Soda Trucks

Materials

soda truck pictures (page 49)
cardstock
crayons or markers
scissors
glue
paper grocery bags
3" x 5" index cards
clean, empty soda cans

Directions

1. Copy two soda truck pictures on cardstock for every paper bag you have. Color and cut apart cans and trucks.

2. Glue a truck picture to each grocery bag.

3. Print a two-digit number on an index card and place it in the chart.

4. Count real soda cans to match the number on the index card. For each can, have a student place a can picture in the chart. When ten cans are counted, place them in a grocery bag and replace can pictures with a truck picture. This implies that each truck holds ten cans.

5. As a class, count cans by tens (trucks) and ones (cans).

6. Clear the chart, empty the bags, and repeat the activity for a new number.

Counting Real Objects

Invite students to bring collections of small objects from home. Place objects in a basket with several paper plates. Have students count the objects, placing groups of ten on paper plates. Remind students that objects on paper plates are groups of ten and remaining objects are ones.

Graph Paper Sections

Laminate graph paper (with ½" squares) and cut it into squares of 100, rows of ten, and single squares. Divide the pocket chart into three sections labeled *hundreds, tens,* and *ones.* Have a student randomly place graph-paper hundreds, tens, and ones in the appropriate columns of the chart. Count the total number of hundreds, tens, and ones in the chart, and have the student write the resulting number on the chalkboard.

Toss and Count

Draw ten sections on the inside of a shoe box lid. Label the sections *0–9.* Place the lid, two counters, and a collection of manipulatives on a table. Designate one counter as the tens digit and the other as the ones. To play the game, students toss both counters into the lid, write the two-digit number on a piece of paper, and count manipulatives to match the number on the paper.

LITERATURE

Can You Imagine . . . ? by Beau Gardner (Dodd, Mead)

Count and See by Tana Hoban (Macmillan)

From One to One Hundred by Teri Sloat (Dutton)

How Much Is a Million? by David Schwartz (Lothrop)

Numbers by Henry Pluckrose (Watts)

One Hundred Hungry Ants by Elinor Pinczes (Houghton Mifflin)

The Three Hundred Twenty-Ninth Friend by Marjorie Sharmat (Four Winds)

Zero! Is It Something? Is It Nothing? by Claudia Zaslavsky (Watts)

Stars

Pocket Chart Math Activities © 1996 Creative Teaching Press

Soda Truck Pictures

Pocket Chart Math Activities ©1996 Creative Teaching Press

The activities in this chapter introduce students to the concept of telling time. To acquire an accurate sense of time, students need repeated experiences with activities involving reading time, writing time, and matching digital and analog clocks. Understanding time as measured on a clock helps students develop a sense of different intervals of time—seconds, minutes, hours, and days. Take advantage of opportunities that arise throughout the school day to discuss sequence of events and passage of time. The goal of this section is to provide meaningful practice to guide students toward mastery of this challenging concept.

TIME

- (Show an analog clock.) Can anyone tell me the time on this clock?

- (Show a digital clock.) Who can read this time?

- Why is it important to be able to tell time?

- What would happen at school if we didn't have a clock in our classroom?

- What are some specific times you talk about at home? (bedtime, time to get up, supper time)

How Long Does It Take?

Materials

3" x 5" index cards
marker
stopwatch

Directions

1. Prepare "task cards" by writing simple tasks such as *Sing the alphabet*, *Count to 20*, and *Name the students in our class* on index cards.

2. Print each child's name on an index card.

3. Place two task cards in the top of the pocket chart and invite children to estimate which task takes longer.

4. Have each child place his or her name card under the task estimated to take longer.

5. Have volunteers perform the tasks. Time each with a stopwatch.

6. Print times on index cards and place them next to corresponding task cards. Discuss which task took more time and which took less.

7. Move cards to the bottom of the chart. Hand out name cards and begin again with new task cards.

Count Out the Time

Materials

small clock faces (page 55)
marker
large clock face (page 56)
blocks or linking cubes

Directions

1. Draw different times to the hour on small clock faces.

2. Provide each student with a large clock face and twelve blocks.

3. Show the "five o'clock" clock face in the pocket chart. Instruct students to place blocks on numbers *1–5* on their clocks, counting as they go.

4. After all children show five blocks, have them clear their clocks.

5. Show a new time and begin again.

The Day's Events

Materials
sentence strips
marker
large clock face (page 56)
tagboard
scissors
brass fasteners

Directions
1. On sentence strips, print sentences about the times when different events occur. (For example, *At 10:00, we go to the dentist. The football game is at 7:00. Dad gets home at 5:00.*)

2. Cut clock hands from tagboard and attach them to clock faces with brass fasteners. Prepare one clock per child.

3. Give each child a clock with movable hands.

4. Place a sentence strip in the pocket chart and read it aloud.

5. Ask children to show the corresponding time on their clocks.

6. Have students hold up their clocks and read the sentence strip aloud as you point to the words.

7. Repeat steps 4–6 with other sentences.

Variation
Paste each sentence strip to the bottom of a large sheet of construction paper. Invite children to draw a picture to match each strip. Glue a reproducible clock face to the page and have children draw the hands on the clock. Assemble finished work into a class book.

Time Vocabulary

Materials
3" x 5" index cards
marker

Directions
1. Print time words (*clock face, hour hand, minute hand, second hand, digital, standard, alarm, watch, morning, noon, night, o'clock, thirty*) on index cards.

2. Place cards in the pocket chart and read them aloud to the class as you point to the words.

3. Invite students to choose cards from the chart.

4. Have students stand, one at a time, read word cards aloud, use them in sentences, and place them back in the chart.

5. Invite other students to pick cards until all have had a turn.

The football game is at 7:00.

Telling Time—Digital and Standard

Materials
small clock faces (page 55)
paper
clipboards

Directions
1. Reproduce several small clock faces and show a different time (to the hour) on each.

2. Place one clock face in the pocket chart. Read the time aloud and invite a student to write the time on the chalkboard.

3. Place a second clock face in the chart. Invite children to read and write the time.

4. Have children hold up their times.

5. Repeat steps 3 and 4 for other clock faces. After times have been written, read each clock face aloud.

Variation
Try this activity placing digital time cards (printed on 3" x 5" index cards) in the chart, and have students show the time on small clock faces with movable hands.

Time Concentration

Materials
small clock faces (page 55)
3" x 5" index cards
marker

Directions
1. Draw hands on each clock face. Print matching digital times on index cards.

2. Place clocks and cards in the chart, face down.

3. Invite children to turn over a clock face and a card, trying to find a match.

4. If a match is made, have children keep the cards. If not, have children turn them back over. Continue until all matches are found.

EXTENSION ACTIVITIES

How Long Is a Minute?
Using a watch with a second hand, have students estimate the length of a minute. Tell them when you begin timing, and have them raise their hands when they think a minute has elapsed. Then have them estimate how many stars, circles, or happy faces they can draw in one minute. Time them as they draw and compare results with estimations.

Hickory Dickory Dock
Copy the nursery rhyme *Hickory Dickory Dock* on sentence strips. Place the strips in the pocket chart, and have children read the rhyme aloud. Give twelve children clocks showing different times. Recite the rhyme, incorporating a different time each time you say it. (The first time, say, *The clock struck one.* The next time, say, *The clock struck two,* and so on.) As each child's time is mentioned, have him or her place the card in the corresponding line of the rhyme in the pocket chart.

Time To . . .
Read *Time To . . .* by Bruce McMillan. Print time cards to correspond with each page of the book. Hand them out to students. Have students place them in the pocket chart as you read the book again.

Journals
Provide each child with a four-page blank journal. Print a time *(8:00 a.m., 12:00 noon, 4:00 p.m., 7:00 p.m.)* on each page. Invite children to keep track of the time and draw what they do at each time that day. (Family members can help at home.) Have children share their journals with classmates the following day.

LITERATURE

Clocks and How They Go by Gail Gibbons (Crowell)

The Completed Hickory Dickory Dock by Jim Aylesworth (Atheneum)

The Grouchy Ladybug by Eric Carle (HarperCollins)

The Sun's Day by Mordicai Gerstein (HarperCollins)

Tick Tock Clock by Sharon Gordon (Troll)

The Time Song by Rozanne Lanczak Williams (Creative Teaching Press)

Time To . . . by Bruce McMillan (Lothrop)

What Time Is It? by Rozanne Lanczak Williams (Creative Teaching Press)

Small Clock Faces

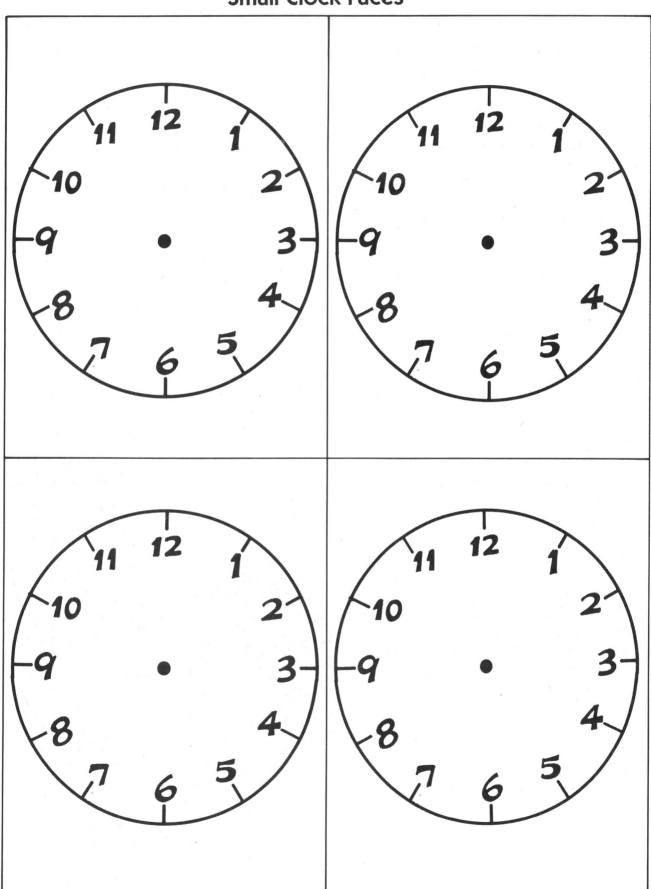

Pocket Chart Math Activities ©1996 Creative Teaching Press

Large Clock Face

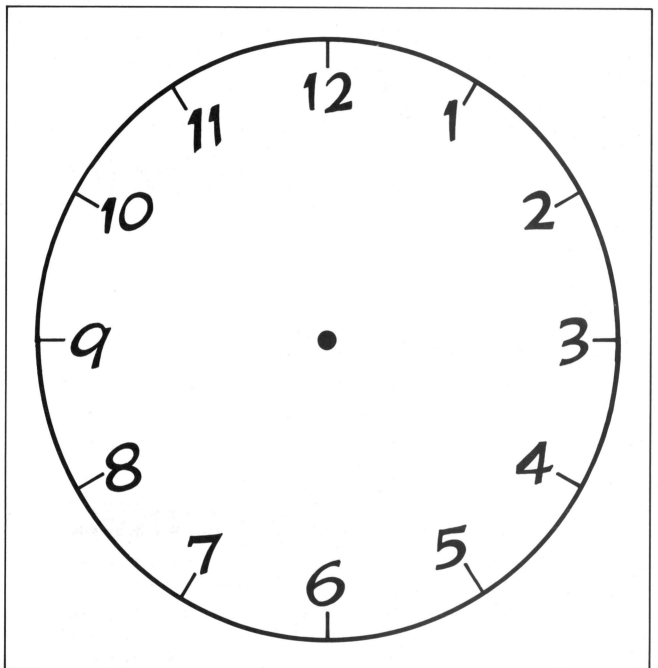

Pocket Chart Math Activities © 1996 Creative Teaching Press

In this chapter, children participate in addition activities that offer opportunities to combine sets of pictures, verbalize findings, read and write vertical and horizontal equations, and match numbers of objects to equation cards. These activities help students realize that joining sets creates equal or larger sets. In order to teach addition and subtraction simultaneously, you may wish to refer to the next chapter as you incorporate the following addition activities.

ADDITION

DISCUSSION STARTERS

- How do we add two groups of objects together?
- When do we use addition in our daily lives?
- If someone gave me two apples, and then gave me two more, how many would I have altogether?
- If there is room for three people at the block center, and two are already there, how could you use addition to figure out how many more can play?
- How are addition and skip counting similar?

Dinosaur Sets

<u>Materials</u>
dinosaur cards (page 62)
crayons or markers
scissors

<u>Directions</u>

1. Reproduce, cut apart, and color several sheets of dinosaur cards.

2. Place two groups of dinosaurs in the top row of the pocket chart. (For example, show a group of three and a group of four.)

3. Invite students to express an appropriate number sentence. *(Three plus four equals seven.)*

4. Have one student combine the sets in the chart and restate the number sentence.

5. Repeat steps 2–4 until all have had a chance to combine dinosaur sets.

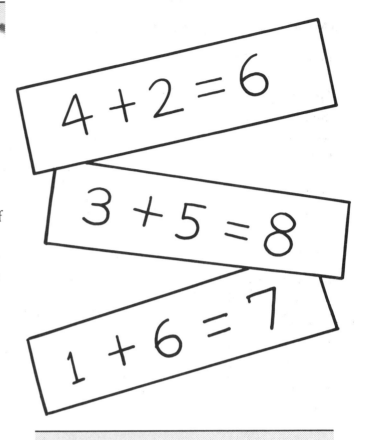

Equation Cards

<u>Materials</u>
marker
sentence strips
paper linking cubes (page 26)
manipulatives (blocks, linking cubes, counters)

<u>Directions</u>

1. Print on sentence strips equations appropriate for students' learning level.

2. Place one equation in the pocket chart, reciting the equation aloud.

3. Have a student place two corresponding sets of paper linking cubes in the chart, combine the sets, and recite the equation aloud.

4. Have the rest of the class use manipulatives to make and combine sets independently.

5. Continue placing new equation cards in the chart, having students form and add sets together.

Equation Matching Game

Materials

dinosaur cards (page 62)
scissors
marker

Directions

1. Cut 20 dinosaur cards in half. Print an equation on one half of each dinosaur. Print the answer to each equation on the other half.

2. Place equation halves face down on the left side of the pocket chart. (Leave the top two rows open.)

3. Place halves with answers face up on the right side of the chart.

4. Invite students to turn over an equation half and find the matching answer.

5. Have children place completed dinosaurs in the top rows of the chart.

Set Combinations

Materials

number cards (numbers 0–9 on index cards)
felt shapes
5" x 7" index cards
marker

Directions

1. Place the 5 card in the top of the pocket chart.

2. Invite students to place combinations of five felt shapes in the chart. For example, students might show two yellow circles and three pink stars, or four blue squares and one green tree.

3. Have students recite number sentences defined by their shape groups.

4. Print an equation card for each combination, modeling both vertical and horizontal equations. Place equation cards in the chart near the sets of felt shapes.

5. Repeat the activity with a different number.

Variation

Invite students to make number combinations using manipulatives such as toothpicks, buttons, or paper clips.

Across the Valley

Materials
dinosaur cards (page 62)

Directions

1. Prepare three dinosaur cards for each of several addition equations. On two cards, write the equation in numerals without the answer. On the third card, show the number of dots that answers the equation.

2. Have children form two lines and sit facing each other.

3. Hand each child a dinosaur equation card. Matching pairs should be in opposite lines. Place dot cards randomly on the floor between the lines.

4. Call out an equation. Have children holding that equation find the dinosaur dot card with the answer.

5. Have them place all three cards in one row of the pocket chart.

6. Continue calling out equations until all cards have been placed in the chart.

Addition Partner Equations

Materials
3" x 5" index cards
marker
2" paper squares

Directions

1. In advance, write several equations on index cards. Print answers to the equations on separate cards.

2. Place an equation card in the top row of the pocket chart, several paper squares in the middle, and various answer cards in the bottom row.

3. Invite a pair of students to complete the equation by counting out the two sets of squares, placing them beneath the equation, and adding. Have them place the answer card at the end of the equation. Invite the rest of the class to work the equation independently.

4. Clear the chart and call another pair to complete the next equation. Play until all have had a turn.

EXTENSION ACTIVITIES

Story Problems

Recite simple addition story problems using dinosaurs as characters. Invite children to place dinosaur cards (page 62) in the pocket chart to correspond with story problems.

Equation Match-Up

Print equations on index cards. Invite children to work with partners near the pocket chart. Have one child place an equation in the chart and the other combine sets of felt shapes to match. Have partners recite and record number sentences.

Dinosaur Art

Have students make dinosaur scenes on 12" x 18" drawing paper. Invite children to draw sets of dinosaurs and write or dictate corresponding dinosaur story problems. Have them write matching equations.

LITERATURE

Addition Annie by Davie Gisler (Childrens Press)

Anno's Counting House by Mitsumasa Anno (Philomel)

Count-a-Saurus by Nancy Blumenthal (Macmillan)

Even Steven by D. Irons (Rigby)

Fish Eyes by Lois Ehlert (Harcourt Brace Jovanovich)

Little Number Stories: Addition by Rozanne Lanczak Williams (Creative Teaching Press)

Pigs Plus: Learning Addition by John Burningham (Viking)

Ten Black Dots by Donald Crews (Greenwillow)

Dinosaur Cards

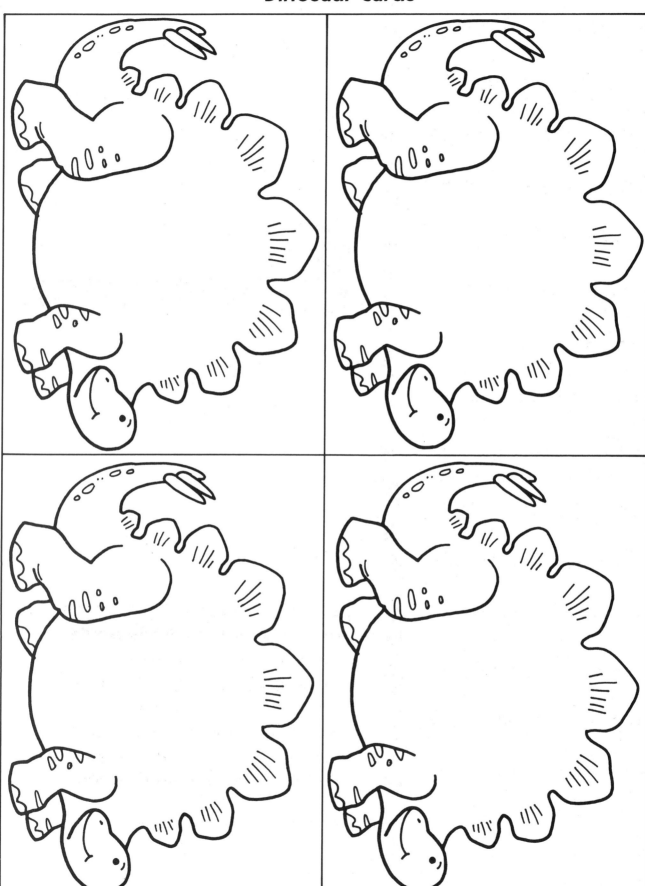

Pocket Chart Math Activities © 1996 Creative Teaching Press

As young children begin to explore the concept of subtraction, they need many experiences separating sets of objects. Such experiences help students realize that separating a set creates equal or smaller sets. Encourage children to verbalize their findings during the following activities to develop subtraction vocabulary (*take away, left over, remaining, minus*). Pocket chart activities in this chapter provide students with opportunities to explore subtraction using sets of objects, modeled equations, and written equations.

SUBTRACTION

DISCUSSION STARTERS

- What do you know about subtraction?
- If I draw three balloons on the chalkboard and erase two of them, how many are left?
- (Write *3 − 2 = 1.*) Who can read this equation?
- Who would like to draw some objects on the chalkboard, take some away, and tell a number sentence to match?
- When do we use subtraction in real life?

Coverup

<u>Materials</u>
stickers
3" x 5" index cards

<u>Directions</u>

1. Apply stickers to index cards so that each card displays a different number.

2. Place a card with four stickers in the chart. Invite the class to count the stickers aloud.

3. Cover one sticker with a blank index card and ask a student to recite an equation to match. *(Four take away one equals three.)*

4. Cover two stickers and ask another student to recite an equation to match. *(Four take away two equals two.)*

5. Continue covering stickers as children recite equations.

6. Use a new sticker card to repeat steps 2–5.

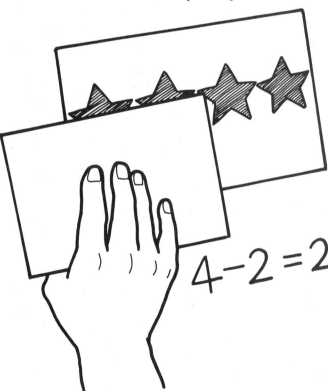

Simple Subtraction Equations

<u>Materials</u>
paper bugs (page 67)
manipulatives

<u>Directions</u>

1. Place several paper bugs in the chart.

2. Take away some bugs and tell students the corresponding subtraction equation.

3. Place a new set of bugs in the chart and remove one or more.

4. Have students work the equation independently with paper bugs or other manipulatives.

5. Ask children to raise their hands when they have the answer to the equation.

6. Repeat for a variety of equations.

Story Problem Envelopes

Materials

story cards (page 68)
crayons or markers
scissors
glue
large envelopes
paper bugs (page 67)

Directions

1. Reproduce, color, and cut apart story cards. Glue each picture to an envelope.

2. Place envelopes and several paper bugs in the pocket chart.

3. Recite a simple subtraction story based on a story card, placing bugs in the appropriate envelope. (For example, *Five bugs were flying near a flower. A boy caught three of them in a jar. Two bugs were left.* Place three bugs in the envelope. Leave two in the pocket chart.)

4. Have a student work a second problem. Guide him or her in removing the designated number of bugs from the row and placing them in the corresponding envelope.

5. Repeat steps 2–4 until several students have participated.

Bug Equation Match-Up

Materials

paper bugs (page 67)
marker

Directions

1. Print subtraction equations on paper bugs.

2. Create matching dot cards by drawing dots on paper bugs and crossing off the number to be taken away. Make a dot card for each equation.

3. Give each child an equation card and place dot cards in the chart.

4. Invite each student to read his or her equation card aloud and place it in the chart with the matching dot card.

5. Play until all cards have been matched.

Domino Subtraction

Materials

domino patterns (page 69)
marker
3" x 5" index cards

Directions

1. Draw different numbers of dots on domino patterns and laminate for durability.

2. Place dominos in the top row of the pocket chart.

3. Demonstrate using dominos as subtraction equations by covering one set of dots. For example, a domino with six dots on one side and two on the other creates two equations: $8 - 6 = 2$ and $8 - 2 = 6$.

4. Have students repeat step 3 for other dominos. Write equations on cards and place them in the pocket chart as students cover dominos.

EXTENSION ACTIVITIES

Domino Grab Bag

Have children work in pairs. One child pulls a domino from a bag while the other recites corresponding subtraction equations. Invite partners to fill in blank domino patterns and write corresponding numerical equations below.

Spider Web Stories

Invite each child to draw a web on white paper. Have children draw large spiders on their webs. Send webs and paper bugs home so students can create and share story problems with family members. As a special treat, invite students to take home the *Spiders, Spiders Everywhere!* floor puzzle by Creative Teaching Press (CTP 4212). Suggest that they use the pull-out pieces to create subtraction stories.

Hide-a-Bug

Print a subtraction equation on the bottom of ten or more small paper cups. Place plastic bugs in a shoe box. Invite children, one at a time, to place cups on the table with the equation showing. Instruct them to count bugs for the beginning of each equation and take away the number shown by covering them with the cup. Have children ask friends to check their work.

LITERATURE

Five Little Monkeys Jumping on the Bed by Eileen Christelow (Clarion)

How Many Snails? by Paul Giganti (Greenwillow)

Little Number Stories: Subtraction by Rozanne Lanczak Williams (Creative Teaching Press)

One More and One Less by Giulio Maestro (Crown)

The Shopping Basket by John Burningham (Crowell)

Take Away Monsters by Colin Hawkins (Putnam)

Ten Monsters in a Bed by Rozanne Lanczak Williams (Creative Teaching Press)

Who Took the Cookies from the Cookie Jar? by Rozanne Lanczak Williams (Creative Teaching Press)

Paper Bugs

Pocket Chart Math Activities © 1996 Creative Teaching Press

Story Cards

Pocket Chart Math Activities © 1996 Creative Teaching Press

Domino Patterns

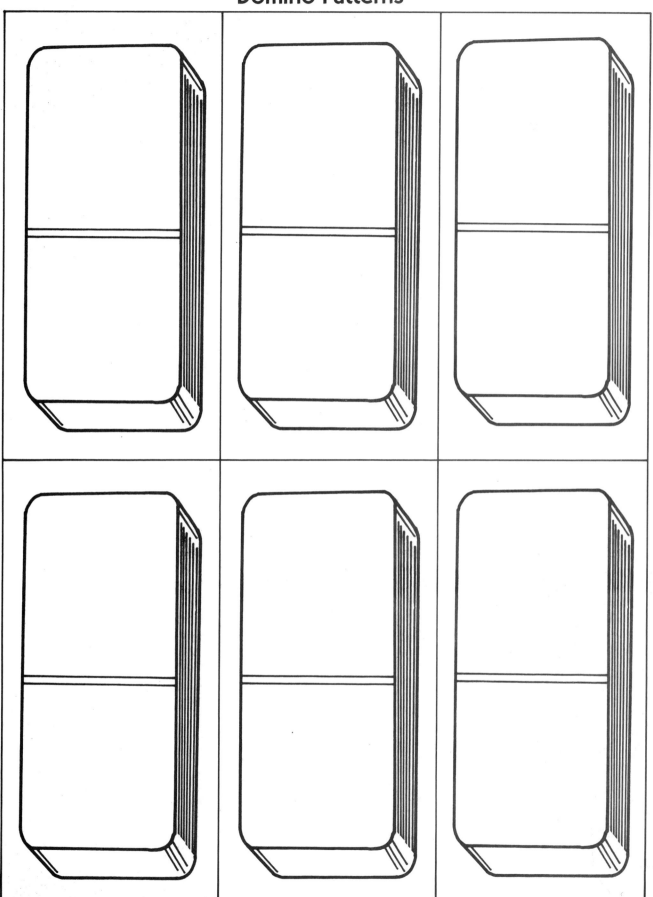

Pocket Chart Math Activities ©1996 Creative Teaching Press

Before children can understand the concept of fractions, they must first think of fractions as parts of a whole which can be separated and reassembled to form the same whole (conservation of a whole). Children must also understand that fractional parts must be equal in size. The activities in this chapter help children experience fractions as equal parts of a whole and as equal parts of a set. As children participate in fraction activities, they learn to relate fractions to real-life experiences.

FRACTIONS

DISCUSSION STARTERS

- Have you ever shared a pizza with others and cut it into equal parts?
- Who can cut this paper plate in half?
- Who can draw two lines on this square to divide it into four equal parts?
- Can someone divide this bag of buttons into two equal piles?
- If I want only half a glass of milk, how much should I pour into this cup?

Shape Fractions Match-Up

Materials
3" x 5" index cards
crayons or markers
shape fractions (page 74)
scissors

Directions

1. Prepare fraction cards by printing $\frac{1}{2}$, $\frac{1}{3}$, and $\frac{1}{4}$ on index cards.

2. Reproduce, color, and cut apart shape fractions.

3. Place shape fractions in the top of the pocket chart and fraction cards in the bottom.

4. Invite a student to place a fraction card next to the corresponding shape fraction. Encourage the student to tell what fraction is colored.

5. Repeat step 4 until all fractions are matched.

Equal Sections

Materials
3" construction-paper squares
linking cubes

Directions

1. Place twelve paper squares in a row of the pocket chart to represent a stack of linking cubes.

2. Invite each student to stack twelve linking cubes.

3. Divide the paper stack in half by separating the first six squares from the second six. Have students do the same with their cubes.

4. Reconnect the stack. Ask students, *How can we divide our stack into thirds or three equal sections?* When children respond and show how to divide their stacks, do the same with the squares in the chart. Repeat for fourths and sixths.

5. Repeat the activity with a different number of cubes. (Stacks of eight, sixteen, and twenty-four work well.)

6. Have students record their fractions by laying paper squares in three "stacks" of twelve on construction paper. Have them divide one in half, one into thirds, and one into fourths, and glue them down.

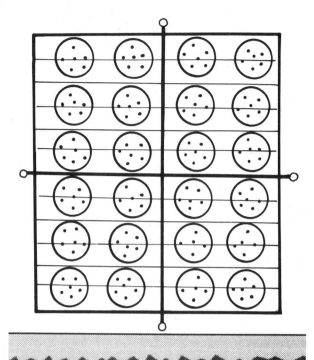

Food Hunt

Materials

food fractions (page 75)
scissors
crayons or markers

Directions

1. Reproduce, color, and cut apart food fractions. Be sure to have one picture for each child.

2. Distribute pictures randomly to students. Have students find others who hold the remaining portions of their food and lay their pieces together on the floor to form a whole.

3. Call out a food and have students name the fractions they are holding and place pictures in the pocket chart to form a whole.

Variation

Prepare food cards by cutting pictures from magazines and gluing them to 5" x 7" index cards. Cut each picture into halves, thirds, or fourths. Repeat the activity using the new picture cards.

Cookie Time

Materials

24 three-inch construction-paper circles
yarn

Directions

1. Place "cookies" (paper circles) in the pocket chart. Count the cookies aloud as you remove them from the chart and place them in a pile on the floor or table.

2. Hang a piece of yarn from the center of the pocket chart, dividing the chart in half.

3. Invite students to divide the group of cookies in half, placing twelve cookies on each side of the pocket chart.

4. Remove the cookies, attach a yarn strand to divide the chart into quadrants, and invite other students to divide the cookies into four equal portions. Repeat for thirds, dividing the chart into three sections by rearranging the yarn.

Variation

Have the rest of the class use manipulatives, paper, and yarn to repeat the activity individually. Have them record their results by drawing or gluing circles on construction paper.

Food Fractions

Bring in foods that divide easily into fractions, such as apples, tortillas, graham crackers, pita bread, and oranges. Invite children to divide foods into fractions before munching away.

Literature Link

Read *Lunch with Cat and Dog* by Rozanne Lanczak Williams. Have children draw foods from the story and write corresponding fractions. Combine student pages into booklets and encourage children to retell the story. Invite students to work with the *Lunch with Cat and Dog* floor puzzle by Creative Teaching Press (CTP 4208).

Fraction of a Set

Prepare plastic bags of manipulatives containing eight, nine, twelve, sixteen, and twenty-four objects each. Encourage children to work in pairs, dividing sets of manipulatives into equal shares and experimenting with halves, thirds, and fourths. Have children place equal shares in plastic cups.

Birdseed

Fill a small bucket with birdseed. Mark $\frac{1}{2}$, $\frac{1}{3}$, and $\frac{1}{4}$ on the sides of several different clear plastic containers. Invite children to explore fractions by filling containers one-half, one-third, or one-fourth full. Provide a set of measuring cups and invite children to discover how many one-half, one-third, and one-fourth cups fill each container.

The Doorbell Rang by Pat Hutchins (Greenwillow)

Eating Fractions by Bruce McMillan (Scholastic)

Fractions Are Parts of Things by J. Richard Dennis (HarperCollins)

Gator Pie by Louise Mathews (Sundance)

Half and Half by J. Nelson (Modern Curriculum Press)

How Many Ways Can You Cut a Pie? by Jane Belk Moncure (Child's World)

Lunch with Cat and Dog by Rozanne Lanczak Williams (Creative Teaching Press)

The Philharmonic Gets Dressed by Karla Kuskin (HarperCollins)

Shape Fractions

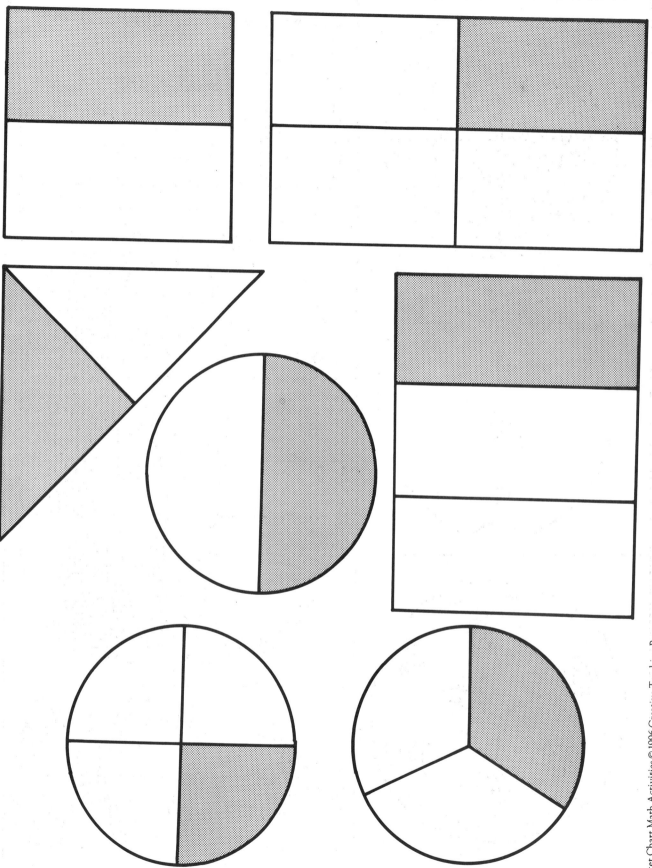

Pocket Chart Math Activities © 1996 Creative Teaching Press

Food Fractions

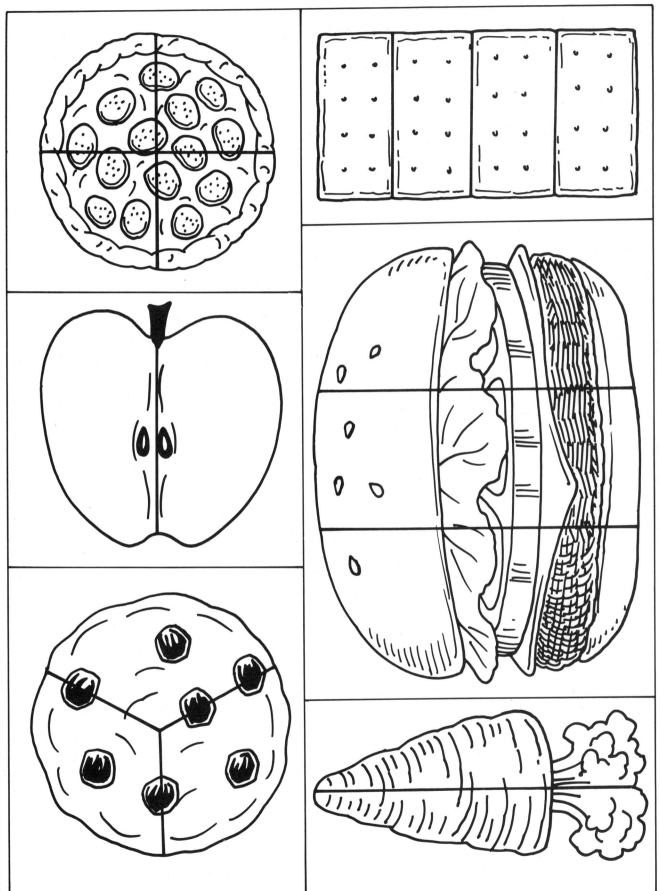

Pocket Chart Math Activities ©1996 Creative Teaching Press

Young children enjoy learning about measurement, especially when it involves objects and situations from their world. The pocket chart activities in this chapter provide students with opportunities to measure the length, mass, and capacity of common objects. Children may need some assistance when first using measuring tools such as scales, rulers, and measuring cups, but invite them to freely explore non-standard tools such as yarn and paper clips. Encourage children to estimate before they measure and make comparisons of measurements they record. Invite them to verbalize their findings as they participate in the activities.

MEASUREMENT

DISCUSSION STARTERS

- What tools can we use to measure ourselves?
- If we had to measure something short, like a book, what might we use?
- How can we measure the weight of an apple?
- Is there a way to measure how much sand two different-sized buckets hold?
- When we cook, bake, or build something, how do we use measurement?
- Why is knowing how to measure things useful and important?

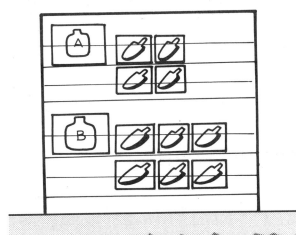

Beach Ball

Materials
beach ball
masking tape
string
scissors
3" x 5" index cards
marker

Directions

1. Have children sit in a circle. Pass the beach ball around the circle so children can observe its size.

2. Wrap masking tape around the circumference of the ball.

3. Have each child estimate and cut a string the length of the circumference of the ball.

4. Have children test the accuracy of their estimations by wrapping their strings around the circumference of the ball.

5. Have each child tape his or her string to an index card. Arrange cards in the chart from longest to shortest to show the range of estimations.

6. Place index cards reading *too long, too short,* and *just right* in the chart, and invite children to place their estimation cards under the corresponding heading.

Scoops

Materials
5 different-sized jars
5" x 7" index cards
marker
scissors
plastic scoop
sand
paper scoops (page 80)

Directions

1. Draw a picture of each jar on an index card. Label jars and pictures A, B, C, D, and E.

2. Reproduce and cut apart paper scoops.

3. Have one student count scoops as he or she fills jar A with sand.

4. Have a second child count the same number of paper scoops and place them in the chart next to the picture of jar A.

5. Continue having students count scoops as they fill the remaining jars with sand. Invite other students to place the corresponding number of paper scoops in the chart.

6. Read the completed chart aloud, counting and pointing to paper scoops for jars A–E.

Variation
Record learning by gluing jar pictures and paper scoops to tagboard. Have students fill new jars (of varying shapes and sizes) with water, rice, or birdseed for further exploration.

More or Less

Materials

eight 5" x 7" index cards
marker
common object pictures (page 81)
scissors
real objects to correspond with pictures
balance scale

Directions

1. Print *more* at the top of four index cards and *less* at the top of the rest.

2. Copy and cut apart common object pictures.

3. Hold up two real objects of different weights. Place corresponding picture cards in the pocket chart. Survey students to find which object they think weighs more.

4. Invite a child to place the two items on the balance scale.

5. Place the *more* and *less* cards in the pocket chart above the appropriate picture cards.

6. Compare weights of other objects and record *more* or *less* in the pocket chart.

Variation

Use a kitchen scale to measure and record the number of ounces or grams each item weighs. Print weights on word cards and place them in the pocket chart next to the picture cards. Invite children to mix and match picture and weight cards as they weigh the items again.

Longer Than, Shorter Than

Materials

picture strips (page 82)
crayons or markers
scissors

Directions

1. Copy, color, and cut apart picture strips.

2. Place picture strips in the bottom half of the pocket chart in random order.

3. Invite a student to place two strips in the left side of the pocket chart. Have him or her recite a "longer-than/shorter-than" sentence. For example, *The necklace is shorter than the skateboard.* (Be sure children place pictures against the left side of the pocket chart to obtain an accurate comparison.)

4. Continue as other children compare measurements of different picture strips.

Variation

Invite children to draw or cut pictures from magazines of other items to use with this activity. Or, cut yarn pieces of different lengths and colors and tape to strips of tagboard. Have children use these in place of picture strips.

Estimating Weight

Place a collection of common objects, a balance scale, and recording paper at a table. Have students choose two objects, estimate which weighs more, and draw the objects on a recording page. After comparing the actual weights, have children circle the item that weighed more.

Estimating Volume

Fill a tub with sand. Toss in a variety of sand pails, shovels, and scoops of varying sizes. Invite children to estimate the number of scoops that will fill each pail, then check their estimations.

Estimating Length

Tape varying lengths of yarn to strips of tagboard. Prepare two of each length. Invite children to estimate which strings are the same length and place matching pairs together.

Rice, Water, or Birdseed?

Place buckets of rice, water, and birdseed on a table. Provide scoops and containers in a variety of sizes and shapes. Encourage children to count and compare scoops as they fill the containers. Have children record their findings in a blank book and share it with a friend.

Equal to My Ruler

Invite children to hold rulers and estimate three items in the classroom that are the same length. Have children draw their guesses on paper and compare the ruler with the selected objects. Invite children to record the correct length of each item.

Who's Taller?

Invite children to sit in pairs on the floor and estimate which partner is taller. Have children stand and compare heights. Invite children to paint a picture of themselves together. Then, have them measure themselves and record their true heights on the painting.

How Big Is a Foot? by Rolf Myller (Bantam Doubleday)

How Tall Are You? by JoAnne Nelson (McClanahan)

Inch by Inch by Leo Lionni (Astor-Honor)

Is It Larger? Is It Smaller? by Tana Hoban (Greenwillow)

Let's Measure It! by Luella Connelly (Creative Teaching Press)

Measuring by Richard Allington and Kathleen Krull (Raintree)

Much Bigger Than Martin by Steven Kellogg (Dial)

The Time Song by Rozanne Lanczak Williams (Creative Teaching Press)

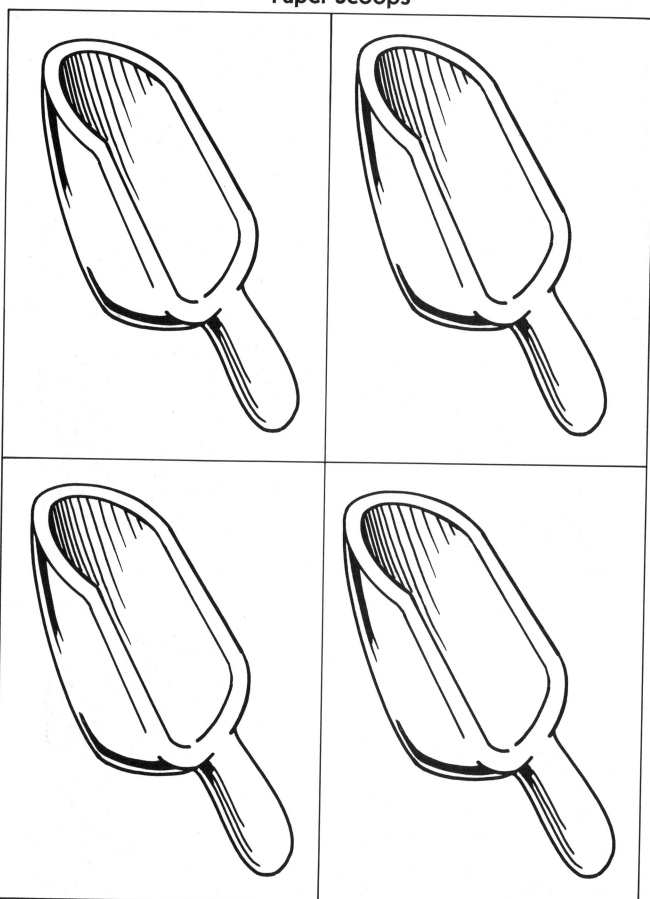

Pocket Chart Math Activities © 1996 Creative Teaching Press

Pocket Chart Math Activities ©1996 Creative Teaching Press

Picture Strips

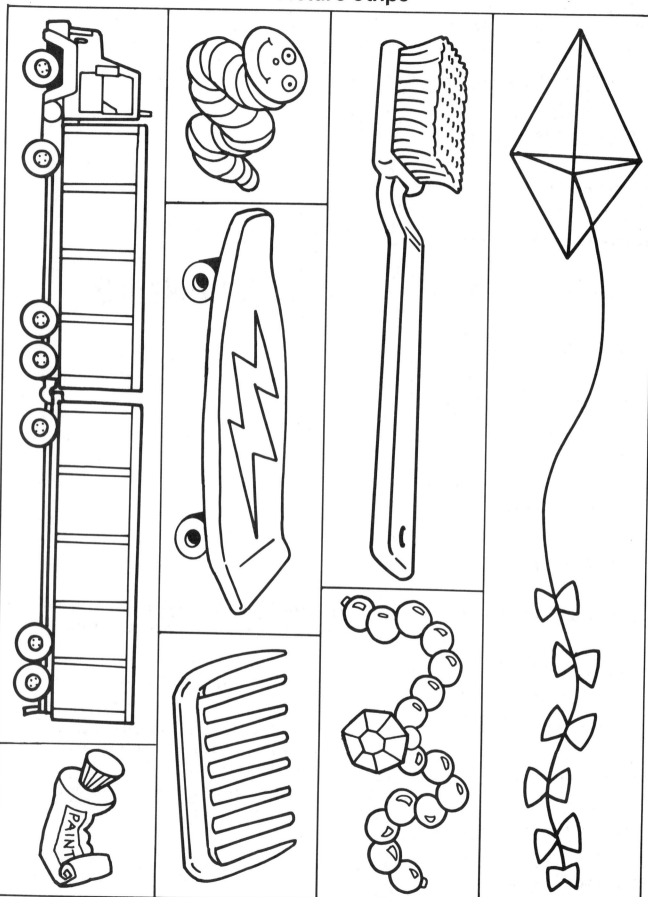

Pocket Chart Math Activities © 1996 Creative Teaching Press

Making and reading graphs is an important problem-solving tool. The pocket chart is excellent for teaching these skills because it allows you to create mobile, manipulative graphs. The following activities help students organize data, record information, and interpret information on a graph. Children become critical thinkers as they acquire mathematical skills such as counting, comparing, recognizing numbers, adding, and subtracting through graphing experiences. Children also develop math vocabulary through repeated graphing experiences as they verbalize, write about, and read information related to graphs.

GRAPHING

DISCUSSION STARTERS

- Who can tell me what a graph is?
- Are there any graphs in our classroom?
- (Show a pile of crayons and pencils.) How can I tell how many of each I have? (Sort the items and display in graph formation.) How does this make it easier to see what I have?
- What can graphs show us?

Question of the Week

<u>Materials</u>

student photos
glue
tagboard
marker
clear self-adhesive paper
sentence strips
5" x 7" index cards

<u>Directions</u>

1. Have students glue their photos to tagboard and print their names at the bottom. Laminate or cover with clear self-adhesive paper for durability.

2. Each week, print a graphing question on a sentence strip and place it in the top of the pocket chart. On index cards, print choices and draw pictures that answer the question. Place answer choices down the left side of the chart. Use questions pertaining to themes being studied, holidays, or special events, such as:

   What is your favorite school subject?
   What is your favorite holiday?
   How many people are in your family?
   What are you wearing?
   What is your favorite sport/game?

3. Create a graph by inviting children to place their pictures next to cards that represent their choices.

4. Discuss the graph by asking questions such as *Which row has the most? Which has the least? Are any equal?*

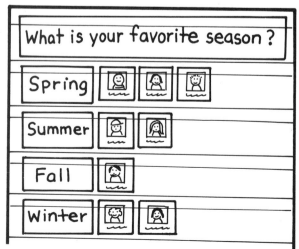

Literature Graph

<u>Materials</u>

book pattern (page 87)
tagboard
glue
crayons or markers
4" x 6" index cards

<u>Directions</u>

1. Photocopy or draw the covers of several familiar literature selections on book patterns and glue to tagboard.

2. Place "book covers" down the left side of the pocket chart.

3. Ask students which book they like most.

4. Have each child fold an index card in half and, with the fold at the top, draw the cover of his or her favorite book. On the inside, have children draw their favorite characters or parts of the story. Encourage children to print or dictate words about their drawings.

5. Have children place their "books" (with the top flap hanging over the pocket) in the corresponding row of the pocket chart and tell why it is their favorite.

6. Discuss the graphed results. Leave the graph in the pocket chart, and invite children to lift flaps and read each other's books.

Weather Graphing

Materials
weather tally sheet (page 88)
weather picture cards (page 89)
scissors
crayons or markers
resealable plastic bags

Directions
1. Hang a weather tally sheet near the calendar. Each day, place one tally mark in the space that describes the day's weather.

2. Copy, color, and cut apart several weather picture card pages.

3. On the last day of the month, ask a child to count the tallies for sunny days. Have the same child place an equal number of sun pictures in the top row of the pocket chart.

4. Repeat for cloudy, windy, rainy, snowy, and foggy days, forming a weather graph in the pocket chart.

5. Count the weather cards in each row and compare numbers.

6. Store cards and tally sheets in resealable bags. Invite children to form graphs during free time.

Variation
Invite children to create their own one-week weather graph by pasting weather cards to 12" x 18" construction paper.

Graphing with a Friend

Materials
plastic scoop
pattern blocks
paper pattern blocks (page 9)

Directions
1. Have one child take an assortment of pattern blocks and sort them in graph formation on a table.

2. Have a second child reproduce the graph by placing paper pattern blocks in the pocket chart.

3. Ask children to discuss what they learned from the graph. Repeat the activity with new blocks.

EXTENSION ACTIVITIES

Graphing Handfuls

Provide an assortment of manipulatives and have children work in small groups. Invite each child to sort and arrange manipulatives in graph formation. Invite children to discuss their graphs.

Snack Time Graph

Incorporate graphing into daily snack time. Offer two or more snack choices and ask students which they prefer. Have them place their snacks on paper plates. Lay the plates in graph formation along a large table. Do the same with beverages, if two choices are available. Discuss the graphs before munching away.

More Edible Graphs

Bring in a snack mix of peanuts, raisins, chocolate chips, and sunflower seeds, or an assortment of dry cereals. Provide a scoop of the mix for each child. Have children sort pieces in graph formation. Then, invite them to recreate their graphs using crayons and graph paper. Invite children to eat their mix to complete a healthy math activity.

Linking Cubes Graph

Provide each child with one linking cube and ask a graphing question such as *Are you eating hot lunch or cold lunch today?* Label two index cards *hot lunch* and *cold lunch*, and have each student stack his or her cube behind the appropriate card. Encourage children to reproduce the graph in the pocket chart using construction-paper squares and similar index-card headings.

LITERATURE

Anno's Counting Book by Mitsumasa Anno (HarperCollins)

Caps for Sale by Esphyr Siobodkina (HarperCollins)

Dad's Diet by Barbara Comber (Ashton Scholastic)

Is It Rough? Is It Smooth? Is It Shiny? by Tana Hoban (Greenwillow)

Mr. Archimedes' Bath by Pamela Allen (HarperCollins)

A Three-Hat Day by Laura Geringer (HarperCollins)

The Very Busy Spider by Eric Carle (Putnam)

We Can Make Graphs by Rozanne Lanczak Williams (Creative Teaching Press)

Book Pattern

Pocket Chart Math Activities © 1996 Creative Teaching Press

Weather Tally Sheet

snowy	cloudy
sunny	windy
rainy	foggy

Pocket Chart Math Activities ©1996 Creative Teaching Press

Weather Picture Cards

Pocket Chart Math Activities ©1996 Creative Teaching Press

The purpose of this chapter is to help students identify and learn the value of pennies, nickels, dimes, and quarters. A number of the following activities invite children to count coins or groups of coins. Activities in this chapter also provide children with opportunities to use real money and transfer their learning to real-life experiences.

DISCUSSION STARTERS

- What do you know about money?
- (Hold up four different coins.) What are the names of these coins?
- What is the value of each coin?
- When have you had to use money?
- When will you need to use money when you're older?

Go Together

Materials
coin picture cards (page 94)
sentence strips
5" x 7" index cards
glue

Directions
1. Glue combinations of coin pictures to sentence strips according to students' level of understanding.

2. Print a matching card with the total value of the coins.

3. Place cards and sentence strips randomly in the bottom half of the pocket chart.

4. Invite students to find matching cards and strips, and place them in the top half of the pocket chart.

5. Play until all cards and strips have been matched. Mix them up and play again.

Variation
Play this game using number words *(one, five, ten, twenty-five, one dollar)* and corresponding picture cards. Invite children to read the coin vocabulary and match pictures with words.

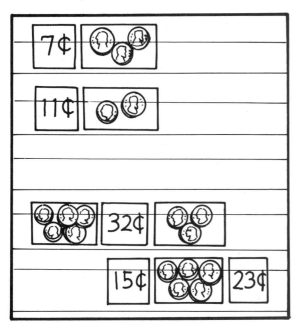

How Much Does It Cost?

Materials
coin picture cards (page 94)
price cards (page 95)

Directions
1. Place four different coin picture cards in the top row of the pocket chart.

2. Review each coin's name and value.

3. Write a price on each price card (*1, 5, 10,* or *25 cents*).

4. Invite a child to place a price card in the chart below the corresponding coin card.

5. Repeat with other price cards until all have been sorted.

Identifying and Sorting Coins

Materials
coin picture cards (page 94)
scissors
coins

Directions
1. Copy and cut apart several pages of coin picture cards.

2. Provide children with collections of real coins.

3. Place one of each coin picture card in the pocket chart. Point to the cards and have students name the coins and their values. Have students hold up real coins as you name each coin in the chart.

4. Place several of each coin picture card randomly in the chart.

5. Invite two students to sort the cards in the chart. Have students count how many of each were sorted.

Coin Combinations

Materials
coin picture cards (page 94)
price cards (page 95)
marker

Directions
1. Write a price on each price card appropriate for students' level of understanding.

2. Place an assortment of coin cards in the bottom two rows of the pocket chart.

3. Invite a student to select a price card and place it in the chart.

4. Ask the child to "buy" the item by placing the correct combination of coin pictures next to the price card.

5. After showing the correct coin combination, ask students if any other coin combinations could be used to buy the item.

6. Remove the price card, place coin cards at the bottom of the chart, and repeat steps 3–5 until all cards have been "bought."

Variation
Include pictures of items cut from magazines and glued onto 4" x 6" index cards.

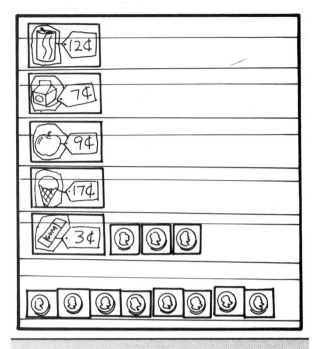

Grocery Shopping with Pennies

Materials
coin picture cards (page 94)
envelopes
food pictures (cut from magazines)
glue
5" x 7" index cards

Directions
1. Place 10–20 penny cards in each "wallet" (envelope).

2. Glue food pictures to index cards and price each between 1 and 20 cents.

3. Place food pictures in the left side of the pocket chart.

4. Give students wallets with which to go "shopping." Have them choose a food, read how much it costs, and place that many pennies next to it.

5. Play until children place the correct number of pennies next to all food pictures.

Variation
When children become more proficient with coins, include nickel and dime cards in the wallets.

A Chair for My Mother

Read *A Chair for My Mother* by Vera Williams. Discuss the story with the class and record children's thoughts on sentence strips. Ask children if they have had similar experiences saving money. Place strips in the chart and invite students to read them aloud while another student shows pictures from the book.

Coin Manipulatives

Prepare coin manipulatives by placing real coins on two-inch tagboard squares and covering them with clear self-adhesive paper. Use these in the pocket chart as children review activities.

What Would I Buy?

Provide each student with a four-page blank book and coin picture cards (page 94). Have students color, cut apart, and paste a card on each page of their books. Encourage children to draw and write or dictate what they would buy with each coin.

Going Shopping

Take a walk to a nearby store when you are in need of classroom supplies. Upon returning to the classroom, write a language-experience story on sentence strips about what you bought and how much each item cost. Give each student a photocopy of the story. Invite students to cut apart sentences, glue them to separate pieces of construction paper, and illustrate each one. Staple pages to create student-made books.

Alexander, Who Used to Be Rich Last Sunday by Judith Viorst (Macmillan)

A Chair for My Mother by Vera Williams (Greenwillow)

Dollars and Cents for Harriet by Betsy Maestro (Crown)

The Hundred-Penny Box by Sharon Mathis (Viking)

The Magic Money Box by Rozanne Lanczak Williams (Creative Teaching Press)

Our Garage Sale by Anne Rockwell (Greenwillow)

The Storekeeper by Tracey Pearson (Puffin)

26 Letters and 99 Cents by Tana Hoban (Greenwillow)

Coin Picture Cards

Pocket Chart Math Activities © 1996 Creative Teaching Press

Price Cards

Pocket Chart Math Activities ©1996 Creative Teaching Press

Dear Family,

Over the coming weeks and months, we will be using a pocket chart in our classroom as a tool for learning math skills. The pocket chart activities in which your child will participate are beneficial because they encourage children to interact cooperatively and verbalize their findings, creating a solid foundation for math language development. The more children participate, the more confidence they gain, and, in turn, the more success they experience. As children become actively involved in skill-building activities, they see their successes and get a real sense of how math skills relate to their lives.

From time to time, your child will bring home finished work and activities to complete. Please assist him or her in reviewing work and completing activities. You will find participation in your child's learning experience to be rewarding for both of you.

In preparation for our pocket chart math activities, we are collecting objects to use for sorting, counting, patterning, and other exercises. If you can contribute any of the following items, please send them to school with your child.

- buttons
- cereal
- dice
- index cards
- large resealable plastic bags
- playing cards
- small boxes
- small plastic toys
- stickers
- trail mix

Thank you for your involvement and support.

Sincerely,